PENGU

Hands

Darian Leader is a psychoanalyst and the author of *Introducing Lacan*, *Why Do Women Write More Letters Than They Post?*, *Promises Lovers Make When It Gets Late*, *Freud's Footnotes*, *Stealing the Mona Lisa*, *Why Do People Get Ill?*, co-written with David Corfield, *The New Black*, *What is Madness?* and *Strictly Bipolar*. He practises psychoanalysis in London, and is a founding member of the Centre for Freudian Analysis and Research and a member of the College of Psychoanalysts UK.

HANDS

Darian Leader

PENGUIN BOOKS

PENGUIN BOOKS

UK | USA | Canada | Ireland | Australia
India | New Zealand | South Africa

Penguin Books is part of the Penguin Random House group of companies
whose addresses can be found at global.penguinrandomhouse.com.

First published by Hamish Hamilton 2016
Published in Penguin Books 2017
001

Printed in Great Britain by Clays Ltd, St Ives plc

A CIP catalogue record for this book is available from the British Library

ISBN: 978–0–241–97400–1

www.greenpenguin.co.uk

MIX
Paper from
responsible sources
FSC® C018179

Penguin Random House is committed to a
sustainable future for our business, our readers
and our planet. This book is made from Forest
Stewardship Council® certified paper.

For Jack, Iris and Clem

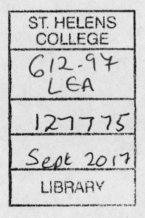

Contents

I

The new era of the Internet, the smartphone and the PC has purportedly had radical effects on who we are and how we relate to each other. The old boundaries of space and time seem collapsed thanks to the digital technology that structures everyday life. We can communicate instantly across both vast and minute distances, Skyping a relative on another continent or texting a classmate sitting at the next table. Videos and photos course through the Web at the touch of a screen, and social media broadcast the minutiae of both public and private lives. In the train, the bus, the café and the car, this is what people are doing, tapping and talking, browsing and clicking, scrolling and swiping.

Philosophers, social theorists, psychologists and anthropologists have all spoken of the new reality that we inhabit today in the twenty-first century as a result of these changes. Relationships are arguably more shallow or more profound, more durable or more transitory, more fragile or more grounded. As the workplace for many becomes virtual, new possibilities emerge to construct lives beyond the nine-to-five paradigm. Whatever we make of these changes, everyone seems to agree that they are indeed changes, that the world is a different place, that the digital era is something incontestably new.

But what if we were to see this chapter in human history through a slightly different lens? What if, rather than focusing on the new promises or discontents of contemporary civilization, we see today's changes as first and foremost changes in what human beings do with their hands? The digital age may have transformed many aspects of our experience, but its most obvious yet neglected feature is that it allows people to keep their hands busy in a variety of unprecedented ways.

The use of hands is indeed changing. The owner of the famous Shakespeare and Company bookshop describes the way that young people now try to turn pages by scrolling them, and Apple have even applied for patents for certain hand gestures. Patent application 7844915, filed in 2007, covered document scrolling and the pinch-to-zoom gesture, while the 2008 application 7479949 covered a range of multitouch gestures. Both were ruled invalid, not because gestures can't be patented but because they were already covered by prior patents.

At the same time, physicians observe massive increases in computer- and phone-related hand problems, as the fingers and wrist are being used for new movements that nothing has prepared them for. Changes to both the hard and soft tissues of the hand itself are predicted as a consequence of this new regime. We will, ultimately, have different hands, in the same way that the structure of the mouth has been altered, it is argued, by the introduction of cutlery, which changed the topography of the bite. The edge-to-edge bite that we used to have up to around 250 years ago became the overbite, with the top incisors hanging over the lower set, thanks to new ways of cutting up food that the table knife made possible. That the body is secondary to the technology here is echoed in the branding of today's products: it is the pad and the phone that are capitalized in the iPad and iPhone rather than the 'I' of the user.

Yet if the way that we employ our hands is changing, the fact that we have to keep our hands busy is nothing new. From weaving to spinning to knitting to texting, human beings have always kept their hands occupied. If the playground parent would once have been knitting or turning the pages of a newspaper, today they are texting and surfing. At home, computer games occupy the hands and fingers, and Minecraft, the world's most successful game, includes a curious onscreen hand that accompanies the player wherever they go. The phenomenal popularity of Lego, likewise, is not just due to clever marketing but to its basic function of giving work to the hands.

Once we recognize the importance of keeping the hands busy, we can start to think about the reasons for this strange necessity. What are the dangers of idle hands? What function does relentless hand activity really have? What role do the hands have for the infant and how does this change during childhood? What links are there between hand and mouth? And what happens when we are prevented from using our hands? The anxious, irritable and even desperate states we might then experience show that keeping the hands busy is not a matter of whimsy or leisure, but touches on something at the heart of our embodied existence.

And this brings us to a paradox that will run throughout the pages to follow. The most obvious answer to the above questions is that we need our hands to do things with. They serve us. They are the instruments of executive action, our tools. They allow us to manipulate the world so that our wishes can be fulfilled. We show our hands to vote, to seal an agreement, to confirm a union, to such an extent that the hand is often used to stand for the human agent that bears it. In zombie and Frankenstein movies, the creatures walk with hands held out in front of them, not to suggest difficulties in vision but, on the contrary, pure purpose.

Yet at the same time, our hands are precisely what disobey. Although there are stories and films where a body part such as an eye, a foot or even an ear becomes animated or possessed, this is nothing compared to the vast number of examples where it is the hand – either joined to the body or severed – that starts to function on its own, and nearly always murderously. When body parts become possessed in horror films, from *The Hands of Orlac* to *Evil Dead*, it is almost always the hands that are controlled by some evil force, rather than the feet, the eyes or the mouth.

In most of these fictions, the hands act in opposition to the conscious volition of the person. They may perpetrate the murder or revenge that the person at some level desires, but which society and their own self-image prohibit. In *The Hand*, the comic-book artist played by Michael Caine has to discover that the carnage wreaked against those who have wronged him is in fact the work of his own severed hand. In other scenarios, the hands may embody the will of someone else entirely, perhaps a spirit or the donor of the surgically grafted member. Whatever the case, they show a division or splitting, as the unhappy hero of these sagas has to fight against their own body.

And don't such divisions persist in everyday life? As we might strive to focus on what our partner or friend is telling us, our hands itch to send a text, to check our email, to update our Facebook page. People complain of being too attached to their phones and tablets and iPads, as if their hands just can't stop touching them. The hand, symbol of human agency and ownership, is also a part of ourselves that escapes us. In what has become one of the most successful cultural products of all time, Disney's *Frozen* explores the dilemma of a girl whose hands do things she doesn't want them to do. Elsa's hands turn whatever they touch into ice, and the story concerns her efforts to censor, control and perhaps accept this part of herself that is, to adapt St Augustine's expression, 'in her more than her'.

When American ex-President Jimmy Carter told the world in 2015 that he was suffering from brain cancer, he said that matters were now 'in God's hands'. To localize agency and volition, it is the hand rather than the mind or brain that is singled out, as if the hands have come to stand for motivation itself. We could think of the final scene in *Terminator* where the cyborg whose only purpose is to eliminate Sarah Connor still inches forward with extended pulverized hand, even after the rest of its body has been destroyed. It is pure purpose, and indeed, in the subsequent films in this series, it is this single preserved hand that will generate the new lethal technologies of the future.

This association of the hand with agency and power has a long history, stretching from biblical texts to Galen, from Calvin's 'manual theology' to Adam Smith's Invisible Hand and beyond. The hand is referred to more than any other body part in the Bible, and appears in the Old Testament more than 2,000 times. In early Christianity, the deity is often pictured as a massive hand emerging from the clouds, and the tablets of the law may originally have been hand-shaped, with five commandments inscribed on each. The ubiquity of this image of the divine hand is sometimes given an iconoclastic interpretation: as it was forbidden to show the face of God, a bodily extremity was chosen instead. The hand appears only because we aren't allowed to see the rest of the body.

The idea of the hand as a tool that serves us to propagate agency is common also in classical times. Where Anaxagoras had argued that humans are intelligent because they have hands, Aristotle, and many after him, countered that they have hands because they are intelligent, as the hands perpetuate our will instrumentally. Medieval etymology had indeed derived the term for 'hand' (*manus*) from that of its 'service' (*munus*). This same idea features heavily in the manuals of rhetoric compiled by Roman orators. They detail complex protocols for the hands to illustrate, punctuate and emphasize speech. Quintilian described the ways that hands could be used to 'request, promise, summon, dismiss, threaten, entreat, show aversion or fear, question and deny', and he claimed to be able to tell just from reading Cicero at which points in his oration the latter would have used gestures when speaking.

Preparing a public speech would have involved careful rehearsal of the splay of the fingers, the angle of the hand and how it would be positioned in relation to the body. Hand gestures formed an integral part of speech, of what Cicero called the 'sermo corporis', the language of the body. These were sometimes considered to be even more important than the content or composition of a speech. As the historian of classical gesture Gregory Aldrete observes, it was presumably no accident that after Cicero's murder it was both his severed head and his hands that went on public display.

This special value accorded to the hands is just as present today as it was in classical times. It seems as if almost every science-fiction, espionage and adventure film contains a scene where the protagonists have to override a computer manually. The sheer frequency of this motif suggests that it touches on something profound for us. We could interpret it as simply human unease about the power that machines have over us, but isn't it also an attempt to sustain the belief in the hand as agency, as the point of ultimate control?

This emphasis on the hands as embodiments of agency is so powerful that the presence of a hand can even be posited to explain what seems to be action without an agent. Neurology describes the 'alien hand' syndrome, in which one hand may act at cross purposes to the other, failing to obey the conscious commands of the patient. An early research paper had hoped to term it 'Dr Strangelove's Syndrome' – after the film in which Peter Sellers has to repeatedly use his left hand to stop the right hand from making a Nazi salute – but the peer reviewers did not like this. 'Alien hand' is often distinguished from 'anarchic hand', where the hand disobeys yet is still experienced as belonging to the body.

The case reports are filled with examples in which the hands seem to be at war with each other: one hand closes a book as the other tries to open it, one lights a cigarette while the other tries to extinguish it, one turns on a TV programme while the other tries to change channels. In more sinister examples, one hand tries to throttle the person while the other hand tries to loosen its grip, one hand forces food into the mouth while the other tries to stop it, or one hand tries to drown the swimming patient while the other tries to restrain it.

In Kurt Goldstein's original 1908 paper, his patient had grabbed her own throat and squeezed with the left hand, saying 'the hand does' rather than 'I am doing'. It had required considerable force to disengage it. She tried to make sense of this traumatic insubordination in terms of there being a 'bad spirit in the hand'. She told Goldstein that 'the hand is not normal; it does what it likes', and that she felt as though two people were present, the hand and herself.

Many strategies to pacify the unruly limb are reported by those suffering from alien hand. Physical restraints like slings have been used, but it seems more common for people to try talking to the hand in the hope that it will start to obey. Goldstein's patient would slap it or beg it to calm down as if speaking to a child: 'My little hand, be quiet,' she would say. In another case, a woman who felt that her hand harboured hostile intentions towards her eventually found peace by treating the alien limb like a child, and it would be difficult not to infer that the mother–child relation itself did not have some part to play in her experience.

Some researchers try to distinguish cases where ownership of the hand is denied from those in which it is recognized, despite its apparent autonomy, yet the distinctions and subsequent attempts to correlate these with localizations of cerebral damage are not always convincing. What is clear is that the intentionality of the alien hand is recognized, even if how this intentionality is understood will vary from case to case. The patients perceive that the hand is doing something goal-directed, but may say, like Goldstein's patient, 'the hand did that' rather than 'I did'.

What is especially fascinating in terms of the question of hands and agency is that in some cases, in trying to explain the bizarre behaviour of the alien hand, the person will evoke a third hand that is controlling the insubordinate one. This 'bad hand', in one case, was causing the patient's left hand to act at cross purposes to the right hand, guiding its movements from above. Rather than appealing to a split brain – that is, to 'another' brain or mind – the explanatory concept is 'another' supernumerary hand.

—

This shows us how the hand cannot be easily withdrawn from ideas of ownership and autonomy. If we mean to do something, a hand must be operative, even if our actual empirical hand is disobeying us. This paradox of manual agency is echoed in the strange transformations of the modern idea of freedom. We are continually encouraged to achieve autonomy by the very agencies that deprive us of this possibility. In *Man of Steel*, the latest version of the Superman story, his father explains that on their planet everyone is born with a predetermined future, but that he wanted to create one being who was free, who was different from the others, who could decide for himself. The circularity is obvious: in this gesture of defiance, freedom itself becomes a mandate. In other words, a fate that is predetermined for his unborn son.

Disney's delightful *Brave* illustrates the same process. The young princess's mother insists on an arranged marriage, totally against her daughter Merida's will. To extricate herself, she casts a spell on her mother, which goes slightly awry: the latter is turned into a fearsome bear that sees the princess as either her ally or as a morsel of prey. At the close of the film, Merida makes an impassioned speech in front of her father and a hall filled with suitors about the importance of free choice and independence. And yet what they don't see is that in the shadows at the back of the hall is the bear, guiding her through her argument. The act of autonomy is in fact an act of ventriloquism.

These examples illuminate the paradox at the heart of modern notions of agency and choice. The obligation to be free and to make our own choices is framed within a network of imperatives that come from without, ordering us to be free. The consequences here are quite clear: the more that autonomy and self-determination are valorized, the more that all basic human activities that fail to come under complete conscious control became pathologized.

And this would explain why the field of so-called 'addictions' is expanding so rapidly. The shopping addictions, sex addictions, Internet addictions and phone addictions that fill the diagnostic marketplace have become seen as addictions because they are apparently not under conscious control. But the real addiction that lies behind them is autonomy addiction: the illusion that we can be fully masters of ourselves. The more we buy into this, the more disorders there will be.

These forces have become so powerful that today even staying alive is seen as a choice, something that we have power over. By devoting ourselves to a healthy lifestyle, eating well and exercising, we will prolong our lifespan. Even when he is being attacked by a vicious monster, Will Smith can still calmly make the free choice of survival. In *After Earth*, he describes how he has been pinned down underwater, with a pincer through his shoulder. He realizes that his time is up and he can see the blood bubbling up through the water. Then it all slows down. Looking at the pincer, 'I decide I don't want that in there any more, and I pull it out and he lets me go.' 'Fear,' he continues, 'is a choice. We're all telling ourselves a story and that day mine changed.'

How different from the cognitive universe of the concentration camp survivor. Where Smith's character made a rational choice at a moment of supreme self-determination, many of those who escaped the camps attributed their survival not to some set of skills but to pure chance. Camp testimonies don't tell us that 'fear is a choice', or that survival depended on working on oneself. Rather, a terrifying contingency was at work, one that has little place in today's world of self-development and realization.

—

The mandate to survive, to live longer, has also introduced into the twenty-first century what we could call a separation of two lives. In earlier times, there was life on earth and then, for many, the belief in another life beyond that. But today we have a new demarcation. It is not between earthly and heavenly life, but between two lives lived on earth, a purely biological one and a real, lived, experiential one. The State in most Western countries now has a duty to keep its citizens alive, and the notion of living has become reduced to basic biological parameters.

The recent case of Ashya King brought this out clearly. After a brain tumour had been diagnosed in this five-year-old boy, his parents wanted to take him abroad for proton beam therapy rather than the standard radiotherapy available in the UK. Proton therapy is used in many other countries to target certain types of cancer, but the UK has not yet built the particle accelerator necessary for it, due to the high costs involved. Although British doctors insisted that Ashya should have the treatment they could offer, King's parents left the country with him, but were soon tracked down and arrested in Spain, as if they had kidnapped their own child. During their seventy-two-hour detention, the five-year-old was separated from his mother and father for the first time since his birth.

Although they were thankfully released following a public outcry, the case shows this clash of two lives. There was the pure biological life of the boy, and then there was the actual, real life he had, including within it his attachment to his parents. One hundred years of biology, psychology, not to speak of psychoanalysis, shows beyond any reasonable doubt that the health of the body is linked to our relationships. As Susie Orbach put it, the single most important discovery of the twentieth century was that hard-wiring and soft-wiring are the same thing. A baby given food and drink yet deprived of any emotional contact with its caregivers will perish. Even Aristotle recognized that without sight or hearing, there could be life, but without touch, death would follow. Yet in the harsh world of modern medicine, although lip service may be paid to the idea, biological life is brutally separated from real life.

In this example, arresting the parents meant treating the biological survival of the boy as entirely separate from the question of what his parents meant to him, and what value being close to them had. He had to be 'treated' in the way that some of his British doctors considered correct. Instead of recognizing an imbrication of biological and emotional life, the two were savagely torn apart, as if a genuine demarcation were possible. We can see the same violence operating in the many children's wards where monitors are scrutinized twenty-four hours a day for any anomaly yet parental visits kept to a minimum. Mothers, indeed, are often referred to here as 'traffic', rather than as, say, 'life-support systems'.

This separation of two lives functions at a less dramatic level in people's everyday customs and habits. Those who can afford it eat blueberries and broccoli, grains and goji, and may spend a great deal of their time going to the gym, jogging and generally keeping fit. When asked why, the answer is obvious: to stay healthy, even if this means that the greater part of one's actual life is given up in order to . . . live longer. Eating correctly and exercising may be experienced as obligations rather than pleasures, so that one life is sacrificed for the purely abstract idea of the other life: the affective is given up for the biological. As Imlac says to Rasselas, 'It seems to me that while you are making the choice of life, you neglect to live.'

———

The contemporary vogue for hand-based crafts replicates many of these same contradictions. We are encouraged to counter our apparent alienation in the excesses of the virtual world by returning to traditional activities such as weaving, knitting, model making, gardening, sculpting and general tinkering. Using our hands to make things is supposed to work against the dematerialized universe that we otherwise inhabit, a remainder or a new efflorescence of the grounding, satisfying bodily techniques of the past.

Yet however laudable and enjoyable such practices may be, they are situated squarely in the very same ideology that they are intended to parry. When we look more closely at the marketing and promotion of globalized brands, they rely on exactly the same strategies: an emphasis on each person's uniqueness, their ability to create, the importance of time for oneself, the continuation of family and folk traditions. However different the coffee may be, the individual laboriously and lovingly hand brewing at home and a vast corporation like Starbucks increasingly share the same set of values.

When people talk about why they practise crafts, the central motifs tend to be exactly those of the modern marketplace: the importance of personal choice, the sense of autonomy, the search for pleasure and a work of self-improvement. The point here isn't that any of these activities are intrinsically good or bad, just that the notion of an antithesis between the individual, focused use of the hands that lifestyle gurus currently preach and the world of mass-produced goods and services is an illusion.

The showcasing of these values in the business world of course hides the violence that runs beneath them, and it is difficult to ignore here how some form of careful and delicate manual activity is so often associated with the cruellest of dictators. We could think here of Hitler's watercolour painting, the precision calligraphy of Chairman Mao, or President Snow's gentle rose pruning in *The Hunger Games*. Individualized manual craft and automated destruction seem strangely allied. And don't we see the same logic in the profiles posted by the CEOs of today's largest multinationals, who, while decimating entire cultures, enjoy model building and pottery?

2

Newborns spend most of their waking hours touching their face and body, with about 20% of their time focused on contact between hand and mouth. When the mouth closes on the nipple, the hand tightens into a fist. The harder the infant sucks, the tighter the grip. When the hand picks up an object, it is moved to the mouth to sample and test. Later, the hand will be able to reach for things independently, and movement will seem to be guided first and foremost by the eyes.

The questions of freedom and agency we explored in the last chapter thus have an odd resonance in infant research. The notion of autonomy is usually discussed within the context of the child's separation from the mother and, later, in the adolescent struggle to find a place in the world. But well before either of these, the first battle for freedom is fought between hand and mouth. During our first year of life, the hand must liberate itself from its dominion by the mouth, a process that the early researchers referred to as the quest for 'autonomy'. And as we will see, perhaps this battle is never entirely won.

Babies are already busy with their hands and fingers *in utero*. Hands will often be drawn to the mouth, as fingers fan, curl, flex and extend. Thumb sucking can start as early as eighteen weeks after conception, and at birth babies often have fingers bruised or swollen from sucking. Their hands are nearly always fisted, unfurling over the first couple of weeks, and equipped with what seemed at first to be certain innate reflexes.

In the so-called 'clutch' or palmar reflex, the hand will close when the palm is stimulated. In contrast, in the digital stretch reflex, when the dorsal surface of the fingers is touched, they stretch out. In the Babkin reflex, pressure on the two palms opens the mouth, something that nurses may exploit when trying to feed an uncooperative infant. In the rooting reflex, the snout turns in the direction of a tactile stimulus and the mouth readies to feed. So in this early battery of reflexes, the two primary responses to stimuli would seem to be sucking and clutching.

Later researchers showed that these reflexes were slightly more complex, as they were not entirely automatic. Piaget discerned attention and interest in his children when he touched their palms, and his twelve-day-old son would stop crying when he placed his finger into the boy's hand. Similarly, the rooting reflex doesn't happen if the infant touches its own face, suggesting a differentiation of self-touch from external touch. Small variations together with increases and decreases in intensity suggest that interactions with the caregivers are involved in these basic reactions.

Whatever we make of these movements, hand and mouth are closely connected here. Babies often suck their fists both before and after feeding, and the hand may be open as the nipple is grasped and then tighten as the feed starts. Film footage shows the contraction of the hand echoing the sucking movements of the mouth, the grip tightening with both hunger and the vigour of the feed. Tiny hands clasp the mother's breast or finger, opening and closing on it rhythmically. When the tongue pushes, the finger presses. Sucking and gripping pressure correlate here, and for René Spitz, the activity of the mouth 'overflows' on to the hand, as the rhythms of sucking and swallowing saturate the hand musculature.

For many of the early researchers, this archaic handgrip was less about manipulation than incorporation. Working with blind children, Selma Fraiberg described what she called the hand's 'morbid alliance with the mouth'. One of the boys would claw her and dig his nails into her skin, but she understood eventually that this was not sadism, just his effort to take her in so she would not go away. The hand here was behaving like a mouth, the fingernails like teeth, and the pinching like biting. Mouthing had been transferred to the hand.

Realizing this, the next time he seized and clawed her, she said, 'You don't have to be afraid. I won't go away.' And he stopped. The behaviour that Fraiberg and the boy's mother had experienced as aggressive was now recognized as terror. The hand was not attacking but enveloping.

———

This alliance of hand and mouth is so powerful that in Arnold Gesell's classic textbook on infant development, the index entry for 'Hands' says 'See Mouth and Hand'. The one seems the vehicle of the other, as the mouth's efforts to retain, capture and explore are perpetuated by the hand. The word 'taste', indeed, originally meant touch, an association still retained by many languages. In Milton's *Paradise Lost*, the two terms are used interchangeably, with Eve's act described at one moment as 'tasting' and at another as 'touching'. The same equation permeates art history, where images of Eve touching the fruit and the 'interdicted Tree' in fact outnumber those of her tasting it in depictions of the Fall.

This compact of mouth and hand is usually contrasted to the later fusion between hand and vision. Early on, it is not the sight but the touch of the breast that produces the sucking reflex, and the baby won't necessarily reach towards things that it sees. The careful work of biologists such as Colwyn Trevarthen shows, in fact, that infants do move their hands directed by sight in the first weeks of life, but it is also clear that the hand only becomes properly bound up with vision some months later. The incomplete development of back and neck muscles means that the baby doesn't have the posture necessary for reaching yet, although experiments employing a special support chair show hand movements characteristic of a twenty-week-old in infants as young as five to eight weeks.

In these first months, the mouth dominates over the eye in its custody of the hand. The baby won't initially grasp what she is looking at, or look at what she is grasping. The hands seem to move in and out of the visual field, without yet being subordinated to vision, and most things grasped are still brought to the mouth. The different sense modalities are all operating here, but the hands have not yet transferred their allegiance to the eyes.

We know from studies of cross-modal perception that throughout this time, information is being transferred from one modality to the other. Piaget had described how when he opened and closed his eyes, his daughter would open and close her mouth, while his son opened and closed his hands. The schema of opening and closing crosses sensory modalities. In a more subtle experiment, Andrew Meltzoff gave one-month-old babies dummies of different textures to suck on, which they were unable to see visually. Then a series of dummies was presented to them in full sight. He found that the sucking rates increased when the baby was looking at the dummy that they had actually sucked. The tongue has very fine sensory discrimination, and information about the dummies' texture had been transferred from the mouth to the eye, but without yet producing any attempt at reaching.

Between three and four months, when the baby's hand brings an object to the mouth, there may now be a little delay to look at it. Vision is starting to shape hand movement, but visually guided reaching still tends to end in mouthing. This is less incorporative and more investigative: the mouth is still the main organ of exploration. During this time, the hands are busy: they corral, flip, slide and revolve objects. The child looks at his hands quite often, and soon, at around four or five months, eye and hand work together to grasp seen objects. It is perhaps no accident that the palmar reflex starts to disappear now. If the mouth is still central in reconnaissance and curiosity, the task is being progressively delegated to the hands.

With the apparent triumph of visually directed reaching at around five or six months, the infant can start to reach for what she sees. But as Jerome Bruner noticed, infants will sometimes close their eyes if there is a difficulty in reaching for an object, as if the hands have to be unharnessed from vision at these moments of impasse. Likewise, they often tilt their mouth towards what they are trying to reach, especially if the hands already have something in them, as Darwin recognized when, as a young collector, he tried to capture a coveted beetle with his mouth as his two hands were already filled with rare specimens.

——

The association between eye and hand is so intractable here that it can warp parental response to the child's efforts and, indeed, their very subjectivity. If the child omits or refuses to follow the direction of a parent's gaze, this can be the cause of a certain anxiety. We expect to find the signs of interest in the world in the infant's eyes, and we may become concerned if there is no indication of selective preference, of the fact that they would rather look at one thing than another. When there is sustained attention to some object, we often anticipate the accompanying action: the hand will stretch out, as if desire consisted in a complicity between the eye and the hand.

But when a baby is born blind, the signs of visual interest are absent. Deprived of this key coordinate, a parent might assume that the child is lacking in both sight and desire. But as those who work with the blind describe, this is due precisely to the dominance of the eye–hand association. If the infant doesn't look at what we expect him to look at, this doesn't mean that there is no desire, just that the motor expression of longing that we read in visual fixation and visually oriented posture is not expressed. If we ascribe subjective states to people through their facial expressions, and particularly through the direction and animation of their gaze, we may find it especially difficult to know how to respond to a child who does not seem to show visual preference.

As Fraiberg and her colleagues found, all they had to do was to encourage parents not to read the face of their child but its hands. The signs of interest that were expected from the eyes were perfectly present, but only once the eye was uncoupled from the hand. When seven-month-old Toni's mother presented her daughter with some of her toys one by one, her face remained immobile. But where her face looked bored, her hands didn't. The fingers scanned each toy, eventually bringing them to the mouth to explore. She clearly valued two of the toys over the others, yet this message was conveyed not by the direction of her eyes but by the movement of her hands.

Many blind children won't orient their face towards the toy in their hands, and since visual inspection is the sign that we read as 'interest', it is easy to assume that none of the toys matter. But although her gaze was turned away, Toni's fingers would explore the crevices of the rattle, the bumps on the soap dish and the bristles of the pastry brush. Similarly, when she dropped a toy, there would be no change in posture or sudden cry, but, as Fraiberg noted, her hands told a different story. They would sweep across the table surface, her fingers searching where the eyes could not.

This 'subtle language' of the hands can be missed because we require the gaze and the hand to function as one single unit. We are supposed to look at what we reach for. But once we separate them, we realize that perhaps the control of the eyes is more limited, and we could think here of the opening scenes of Orson Welles's *F for Fake* in which we first see the visual reactions of several Italian men to Oja Kodar's provocative stride down a busy street in Rome and then a close-up of their hands: nervous, jittery and contorted.

During the second half of the first year, oral perception seems to fade. Not everything is put into the mouth, and infants spend a lot of time exploring objects with their hands while looking at them. Postural change is significant here, as the tonic neck reflex – with the head rotated to one side, one arm extended and the other flexed towards the shoulder – characteristic of the first few months gives way to midline orientation of the head, which makes a more symmetrical movement of the hands possible. The earlier position meant that visual attention was channelled to the extended hand, while midline orientation increases its scope.

The infant can by now transfer things from one hand to another, and can reach and attain objects guided by sight. Eye and hand seem to be working in partnership, with a powerful coordination between them. The reciprocity between hands is explored, often with great interest. By the end of the first year, as they become able to oppose thumb to forefinger, and forefinger to thumb, new forms of dexterity become possible: the child can pry, poke and pluck. Eating with the fingers is easy.

As spoon-feeding starts in the latter part of the first year, observers have noted the way that children very frequently need something to keep their hands busy while eating. The traditional transition from breast to bottle involved the same bifurcation. The history of paediatrics shows the range of devices that have been invented – always by men – to facilitate this, and they invariably take the form of a bottle 'plus'. Hugh Smith's famous 'Bubby Pot', introduced in 1770, looks a bit like a coffee pot with a long spout perforated at the end: a piece of fine rag tied loosely over it helped to strain the milk, but also, crucially, 'it serves the child to play with instead of the nipple'.

Babies cannot feed only with their mouths. The hands have to be used too, and Smith's 'fine rag' is just one of the many supplements that have been introduced in recognition of this. Fingers may stroke, claw or cup the breast or bottle, just as they find their way to the mouth itself during feeding. Yet even when a finger or thumb is being sucked, the other hand may search for something to hold. Freud evoked a 'grasping drive' (*Greiftrieb*) here, which combines the infant's sucking with a tugging activity on a part of oneself or of the mother. This could be the mother's hand, earlobe or, famously, the little bit of rag or cloth that Winnicott called the 'transitional object'.

If the early hand activity here can be seen in terms of the body acting as a mouth, with the hands echoing what the mouth is doing or wants to do, Winnicott thought that these little objects had another function. Emerging in the space between mother and child, the transitional object, he argued, allows a certain transcendence of the immediacy of the relationship: this is an object that is neither her nor me, situated in an in-between space that will form the matrix for creativity and growth.

There is a strange and archaic transaction here. Consider the situation. An infant is feeding, with nipple or bottle in mouth, finger perhaps in mouth as well, and yet they are simultaneously clasping or rubbing some object. How different from a simple model of biological nutrition. We see here what psychoanalysis calls the 'drive', defined broadly speaking as everything that is happening in this scene beyond the level of pure need. But what is the role of the object?

To start thinking about this, we have to consider the fingers first. It is often assumed that after the frustration of not finding nipple or teat, the digit may take its place. But fingers are swiftly included in the actual act of sucking at breast or bottle, to produce what Willi Hoffer called a 'competition' between them. It seems as if the sensations activated by sucking, and perhaps swallowing, are sought by the infant in addition to the actual milk they are associated with. The sensations take on a weight of their own, and so the experience of feeding introduces an original splitting.

At one level, the baby wants to relieve the feeling of hunger or thirst, and at another they want to reproduce the sensations of satisfaction initially linked to it. This search is interminable, and the motif of refinding will mark every aspect of human life. Those who use drugs describe the attempt to refind the original hit, and most of our everyday practices, whether choosing food and drink, having sex, playing sport, smoking, shopping or holidaying, are in the service of this same quest. Freud called it the 'identity of perception', the effort to make one experience match another, to map on to it perfectly, to recreate the original feeling.

People strive on a daily basis to reproduce this, as their demands about food and drink become more and more specific. If a newborn baby gulps down milk, how long will it be before they are asking for a 'very hot medium skinny flat white with chocolate on top'? The coffee bars that now fill our streets specialize in catering for this effusion of personal taste, as each person has a list of requirements that must be met by the hot drink. We could see the multiplication of words as proportional to the difficulty in getting to 'the real thing', which consists not in some ultimate object or goal but simply in the process of matching, of finding 'the same'.

This is where advertising and branding draw their real power from. The delight on a child's face when it sees the image of Thomas the Tank Engine appear on TV or on some merchandise is less due to any interest in the rather dreary world of the island of Sodor than in the fact that the image matches the previous one. Children love recognizing things, and brands – unlike tables and chairs and trees – offer a very precise and delineated image that doesn't change from one moment to the next. Branding allows the possibility of refinding better than anything else.

Now, if the finger or thumb offers the conduit to the promised refinding, it may also include within it traces of the one responsible for the original pleasure, or, indeed, frustration. In her memoir of growing up, Terri Cheney describes her moments of self-harm as if they were attacks on someone else: 'The fact that I was injuring my own body didn't really occur to me. That wasn't how it felt. I was my mother's creation, and every cut spited her, not me.'

We can remember here that all of our bodily functions in infancy are linked to our caregivers, and so the love, rage, pain, despair, pleasure and frustration elicited during our interactions with them may become embedded in the bodily functions and zones themselves. Feeding, nappy changing, potty training and, little by little, sleeping all involve more than just the biological body of the child. They link the mouth, the skin, the genitals, the ears, the eyes and the respiratory system to those who caress us, feed us, clean us, wipe us, look at us, talk to us, read to us and sing to us.

Through such processes, these areas of the body become sites of exchange. Later in life, so many practices – from the laying on of hands to the use of the hand in hypnosis to the healing hand of massage therapies – draw on the power of this connection. Today, indeed, in the era of screen-based medicine, patients are often disappointed that during their consultation the doctor did not touch them. Leaving the surgery with perhaps a prescription produced by a purely verbal correlation between what they complained of and the name of some illness they are deemed to suffer from, the absence of the hand may leave them feeling cheated and deprived.

Yet this connection between bodies is there right from the start. When a baby touches the mother's hand or her face while feeding and then moves this hand to their own mouth, does it carry with it a part of the mother or the experience of her? Babies will often caress their own face while putting their thumb in their mouth, as if they were two people: the one sucking and the one caressing, perhaps adapting this from the mutual play of hands and fingers with the mother. Likewise, actions like tugging or pulling at parts of their own bodies, like the ears, may come just after doing this to the mother.

Piaget observed how when he held out his hand to his three-month-old son, the latter would reach for it, yet when his hand was held in the same pose but out of the child's reach, his son would repeat the same gesture on his own hand. Was he just repeating a motor schema or, at some level, did his own hand become assimilated to his father's?

Infants who have been abandoned may rock themselves in the same way that they were once rocked, as if they act as both parent and child. When a two-year-old was asked why she would put her hand in her nappy, she replied, 'When I've got no one to play with.' The hand, then, comes to embody another person, exactly as we see in the many horror stories and films where it is possessed by an evil force.

This troubled cohabitation of the body is perhaps present in the simplest gaze of the neonate at their own hand. Researchers have always been struck by the way in which babies can stare in fascination at their hands, and efforts to link this to developmental schemas of 'control' have tended to be unsuccessful. As Spitz noticed, the fixed gaze has little immediate effect on the child's ability to use the hand, which occurs only after a certain time lag. It is as if the hand is not included in the body, or that it is something more than the body, an alien presence that, as the horror narratives suggest, is never fully part of ourselves.

This hand embodies a strange otherness for the baby, and it may then come to index the presence of someone else, the parent whose love and care are so bound up with the infant's body and bodily functions. A child may suck its thumb dreamily, but then glare at it angrily, as if it were the thumb of someone else. If we were to see the thumb – no doubt too simplistically – as a substitute for the nipple, or, better, as a hybrid element that includes the other within it, this would mean that during our infancy we have to separate not just from the breast but from our own hands.

To return now to the bit of blanket or rag that the baby holds, we see the emergence of a first way out here. In a space where the child and the mother's body may be confusingly caught up with each other, the transitional object offers a point of consistency: neither her nor me. It offers a way, which will be elaborated throughout most people's lives, to be somewhere else thanks to what we hold in our hands.

3

Why is it that in almost every adventure film ever made there is a scene in which one person holds another dangling by the hand? In most, they succeed in pulling their charge to safety; in some, the person hanging loses their grip and falls into an abyss. The sheer frequency of these scenes is quite staggering, and one might guess that they touch on some aspect of early experience. But which one?

As infants, we might be tossed about or swung around like a plane, but hardly left dangling by the hand. Curiously, this would be perfectly possible, and late-nineteenth-century researchers did it many times, showing how a newborn can support almost its entire body weight from the hand, an ability that we lose after a few months of life. If an adult places their forefinger in the baby's palm, it can be clasped so tightly that the body can be lifted up and held in the dangling position by the finger alone. The handgrip of a newborn can be vice-like, and muscle tone is much higher in the first couple of weeks than later on.

This extraordinary strength was understood as a vestige of the time when newborns had to cling to their mother for survival, as many primates still do. The psychoanalyst Imre Hermann believed that separation from the mother's body occurs too early in humans. Much of our behaviour can then be seen as an effort to recapture or compensate for this premature weaning: the way we sleep snuggled up against one another, the palmar reflex, the erotic value of the hands, the prevalence of sucking behaviour and our tendency to cling on to something in situations of danger.

This latter observation is interesting. When under threat, we do tend to either run or hold on to, even if the latter does not serve any immediate survival purpose. Think also of all the books and films where someone hangs on to a rope, a raft, a tree trunk in order to escape some peril, whether man-made or natural. Hermann's theory might then explain the ubiquitous dangling scene as a focal image of our early separation from the mother's body. We long to be pulled back and are terrified of being dropped.

These ideas were a useful counterpoint to the widespread psychoanalytic emphasis on the mouth as the source of all our early experience. Where the child's approaches to the mother were understood commonly as attempts to incorporate her orally, Hermann stressed the drive to cling on to her, and the role of skin and muscle in this process. Hermann's friend Michael Balint evoked here the many acts such as handshaking and the laying on of hands where it seems as if we are asking for help from the other person or identifying in some way with them.

But this fails to account for the objects we cling to. To see them all as simply substitutes for the mother is unconvincing, and Winnicott's effort to separate those things that represent or embody the mother from those that don't seems more fruitful. We could even see the typical scene in which someone clings on to a plank or tree during some flood or earthquake as an image of precisely this difference: if the natural catastrophe evokes the massive omnipotent maternal power over a newborn, the humble object they cling to is akin to the rag or blanket the child enlists to escape.

—

But why the dangling scene? In the first weeks of life, newborns may briefly pursue the mother's hand when it has been withdrawn, or attempt to restore a contact that has been temporarily broken. This might set a premium on touch, but it fails to explain the acute dependency of the dangling: it is, after all, a matter of life or death. If it is perhaps an image of potential abandonment, why the focus on the hands?

Audiences around the world were reduced to tears in 1965 when, in David Lean's film, Dr Zhivago's daughter tells her uncle of the moment when she lost contact for ever with her father. There were people running everywhere, explosions, and, she says with acute emotion, 'He let go of my hand.' She repeats the words, 'He let go of my hand.' The grip of hands is here the very index of human attachment, and it would surely not be an overstatement to describe her subsequent experience of desertion and loss as an abyss.

We could also observe here that the person being held is not always a loved one or a friend but often an enemy. If they are saved, the antagonism might cease or a debt might begrudgingly be contracted. A life is owed now, which is of course one of the most basic features of our emergence into the world. By being born, we owe our life to our parents, a fact which, when it starts to register, is not always palatable, especially if things haven't gone too well between them. It is perhaps no accident that the image of one person holding another by the hand echoes the cord-like structure of our original umbilical link to the mother.

More rarely, in such scenes, the person doing the holding lets go, as we witness in the recent *Dawn of the Planet of the Apes*. The film charts the conflict between the generous and civilized ape Caesar and the pugnacious and disloyal Koba. At the end of the film, after an explosion, Caesar holds Koba dangling over an abyss. As Koba pleads with him for mercy, reminding him of his own adage 'Ape not kill ape', Caesar releases his grip, sending his enemy falling to his death with the words 'You are not ape.'

The scene sadly belies the leftist ethos that characterized the original series. The first *Apes* films were sensitive studies of discrimination, racism and stigma, and even risked box office failure by depicting the end of the world as a consequence of nuclear idolatry. Here, on the contrary, the old racist distinction between 'them' and 'us' is re-established: Koba is killed because he no longer fits into the category of 'ape', and hence is expendable, just as racist atrocities of both past and present are so often predicated on exclusion from the category of 'human'. It would have been nicer if Caesar had just said, 'Yes, Koba, you are an ape, but since you have done such bad things, I'm still going to kill you.'

The release of the handgrip here is lethal, the very difference between life and death, and we could see it perhaps as the exact opposite of the clenching that we saw was so prevalent in the first months of life, practised by the chorus of hand and mouth. The dangling scenes that are such a staple feature of action and adventure yarns thus matter not because they stage an experience that we have all had but, on the contrary, the possibility of its inversion. They thus come to stand for attachment – figured here in the grip – and loss – figured in the release.

—

We could contrast the dangling scene with what is perhaps one of the commonest fears of childhood: that a hand will grab you from under the bed. If you need the grip of one hand to save you from falling, this is a grip that you have to escape, that brings not life but death. The fear is so prevalent that there is even an episode of *Doctor Who* based around its ubiquity. The Time Lord thinks that since every child in the world believes that a hidden hand will grab them, it must index some truth: that there really is something under the bed.

How curious that the very same action of the hand can generate such different polarities of emotion. We depend on the hand not letting go in the dangling scene, but it is its very tenacity that terrifies in the child's bedroom scene. Are these two sides of the same coin? Do the images separate out what are in fact fused aspects of human touch?

Childhood is filled with attempts to differentiate our feelings. Fairy tales tend to offer neat partitionings of good and evil, and children are usually eager to identify who is the good guy and who is the bad guy. This initial taste for division complexifies over time, so that a good part and a bad part may be located within the same character. If Disney's *Frozen* introduced a romance between Anna and Prince Hans, with the danger situated in sister Elsa's ice-dealing hands, we discover that Hans actually has selfish and even murderous intentions himself. I have listened to many young children trying to explain this, to make sense of a goodie turning bad, whereas for older viewers the contradiction is more bearable.

We might wonder, indeed, if Disney's pun is intentional: if a negative force is first embodied in Elsa's hands, as the story unfolds it becomes localized instead in the handsome Prince Hans. The point of malevolence might move from a body part to a character, but it retains the same phonetic consistency: Hands – Hans. As Elsa learns to use her own powers for the good, there is still a bad Hans.

There is certainly a time in childhood when we have to leave the one we are most attached to, and the seemingly erratic behaviour that this dilemma produces often mystifies the parent. The child will run away from the mother and then cling to them, dispatch them coldly and then beseech them for an embrace. Games like It are perhaps treatments of the same problem of proximity and distance: we run away from the one we long most to be with, and it is significant that the privileged point of both contact and contagion here is the hand.

We could also think of the many narratives in which two people are on the run, bound together by handcuffs. They are nearly always characterized by an initial antipathy – cop and criminal, liberal and bigot, innocent and guilty – and yet as the story moves on, a bond of respect or love is established between them. There is often also a moment when one of the two has to make the choice of whether to free the other party or not. Some danger to both of them appears, and there is a crucial decision to be taken, relying entirely on trust. If they free the other, will they turn their weapon on their liberator or, on the contrary, use it for mutual defence against the approaching danger?

Like the games of It, this plot line is predicated on the image of a union followed by a separation. The union, however, is not chosen but enforced, a fact of circumstance, whereas the separation is an act of choice, and it is in between these two moments that the emotional bond between the characters crystallizes. Having their hands joined and then released is the framework within which the unstable and agonizing fusion of love and hate can – in fantasy – become disentangled.

Later on, classroom crushes may aim to mitigate this same pain of ambivalence: there is just one unique object of adoration. Rather than confront the unsettling and volatile mixture of affection and reproach towards the parents, a purer form of infatuation is created. But beyond and beneath this is always the disturbing alloy of opposed feelings. The same hand that we seek comfort from and depend on is also the hand that we must separate from. If, at one level, we are terrified lest the hand drop us, there is perhaps an even greater terror that it will never let us go. The creature under the bed is not just a monster but, more precisely, a monster that grabs us, with a grip that offers no release. The same grip that we needed so much in our infancy now becomes the symbol not of security but of capture. And yet without this threat, how would we be able to separate from our caregivers?

—

The attention paid to grasping in early life has overshadowed the other manual activity that is learnt much later: the ability to let go. If clasping is initially a reflex action, releasing is not, and must be learnt over the course of at least the first year. Aristotle could write that 'The business of the hands is to take hold and to keep hold of things,' but the emphasis on possession neglects this crucial metamorphosis. When infants manage to build towers by stacking boxes or cubes, the most difficult moment is not holding them and placing them but, on the contrary, letting each one go so that it doesn't topple.

Voluntary release is hardly a given here. To let go of something, the flexors are inhibited, and this is not easy for children during the first six months. To release a cube in tower building, for example, the contact of fingers and opposed thumb has to be broken simultaneously. Dropping is thus an art to be learnt rather than a natural reaction. As well as knowing how to open the hand, the timing has to be correctly judged. In their first efforts at lacing shoes, children often pull out the lace as they withdraw their hand.

It is interesting to observe here that so many of the earliest games children play revolve around the oscillation between clasping and then letting go. In one example, an eleven-month-old boy emptied out a box of cotton reels and was trying to put them back one by one, yet the reel would consistently come out with his hand. He was unable to let it go. Clasping was overriding release. When his sister called him, he looked up and the hand opened, allowing the reel to drop back into the box. Things were smoother for Freud's grandson Ernst, who was able to hold on to the cotton reel, throw it and then pull it back in the famous 'Fort/Da' game.

This eighteen-month-old boy started by taking any small object he could find and throwing it away into a corner or under the bed, emitting a long 'Oooo' as he did so. Freud heard the word 'Fort' ('gone') in this sound, and assumed he was playing 'gone' with them. One day he took a wooden reel with a piece of string tied round it and threw it over the edge of his curtained cot so that it disappeared, once again saying, 'Oooo.' He then pulled the reel out of the cot again with a joyful 'Da' ('there'). The game consisted of a disappearance and a return.

While his mother was out, he found a way of making not only the reel but himself disappear. Using a full-length mirror, he would crouch down in order to make his mirror image 'gone', as the mirror did not quite touch the ground. This was a child, Freud noted, who could tolerate well his mother's absences, and 'gone' was one of his first words. To compensate for her absence, Freud thought, the child was staging the disappearance and, later, the return of objects. Where he was passive, he made himself active: from being the one who was left he had turned himself into the one able to manipulate presence and absence.

But throwing the object away may also, Freud added, have satisfied an impulse for revenge, as if to signify to his mother: 'All right then, go away! I don't need you!' Indeed, a year later when his father was absent due to military service, Ernst would take a toy and hurl it away, shouting, 'Go to the fwont!' Isn't there a difference here between throwing and letting go? Many children are able to push things away before letting go of them, as if the rage or defiance in the first action has to be spent to allow a genuine separation.

—

Language preserves this distinction, as we speak of 'letting go' of something that has troubled us only once we have managed to go beyond our anger or the intensity of our attachment. The fact that we perhaps let go less often than we should shows how difficult this process can be. It is ironic that we are continually instructed in contemporary culture to 'move on' and 'let go', and yet a whole set of more dominant imperatives simultaneously requires us to be even more attached to whatever we do, from a pointless training exercise at work to a gym class.

Job applications today have to be accompanied by a letter of motivation, demonstrating that whatever the tasks that work involves, they will be met with unbridled enthusiasm and energy. Motivation here becomes something to be turned on or off at will, rather than an enduring part of one's relation to one's interests. School and college graduates today are in fact increasingly unlikely to find employment in their chosen field, and motivation itself thus loses any connection it may have had to its childhood sources.

If the infant Frank Lloyd Wright could play with wooden building blocks and then go on to create buildings, a comparable trajectory is generally ruled out today due to the well-known changes in the landscape of employment. At the same time, the worker or job applicant has to show absolute passion for the job they don't want to do. Note how, in advertising and branding, the accent has shifted from the product – *our ice cream is delicious* – to the worker's relation to the product – *we're passionate about ice cream*.

If we meet a salesperson, they smile and bubble over with enthusiasm for their hand-crafted, hand-stirred, hand-flavoured product, and then we learn a few weeks later that they have moved on to another totally different company where they are now showing an overwhelming and all-encompassing passion for . . . another product. Rather than a rhythm of attachment and loss, we have a rhythm of attachment and attachment.

It is obviously no surprise then to find the emergence of depletion and depressive states punctuating the forced buoyancy and positivity required in modern life. With no time or space to 'let go' properly or for feelings of loss to be adequately worked through, they hit us from out of the blue, and we often have no idea where they could have come from.

—

This difficulty in registering loss is reflected in the comparable difficulty of processing not absence but presence. When Freud wrote about the 'identity of perception', he suggested that if a large part of our lives will be spent in searching for the lost object, trying to refind the original, mythic experience of pleasure, this will never be entirely absorbed into the experience of desire. If we strive to refind and reattain something situated outside ourselves, there will also be a remnant of our bodily agitation, a tension that inhabits us, linked to the problem of having our early demands met.

No parent can ever provide the response to their child that will eradicate frustration and pain, and no magic can ever solve the problem of the identity of perception. We are caught from the start of life between the search for the satisfaction of need and the need for the repetition of satisfaction, two quite different things. This generates an itching in the body, a state of internal restlessness that we seek to assuage by importing some further stimulation from outside, which we hope will bring satisfaction. This will take the form, Freud thought, of some kind of bodily 'manipulation' analogous, for example, to sucking.

But – and this is perhaps the crucial point – the manipulation will never just take place on its own: it is at the very moment that Freud evokes sucking that he refers to the 'simultaneous rhythmic tugging' of other parts of one's own or another's body. The search for sensual satisfaction is 'combined' here with the activity of the hands, as if the mouth is never enough on its own to attenuate our agitation. Think, indeed, of erotic life: does anyone kiss without also doing something with their hands? The tugging, rubbing and caressing that Freud observed are almost always present here.

Some forms of sexual practice, in fact, exploit this very principle in order to maximize states of tension. If the person's hands are bound and they are forcibly restrained from using them, the resulting sensation of pressure can become a sexual aim in itself. At the same time, the hands are never entirely passive: they fight and wriggle against the ligature that inhibits them, becoming, for some people, the focal point of the entire exercise in subjection.

This activity of the hands is linked not only to the mouth, but to all our other bodily drives, whether oral, anal, scopic or auditory. The search for satisfaction centring around a part of the body will always have the manual accompaniment described by Freud, whether we are looking or being looked at, speaking or listening, biting or sucking. The unsolvable problem of drive tension means that there will always be an excess in the body, a 'too much', that our bodies – and hands – are perpetually trying to exile. Different historical periods have different ways of naming this attempt to divest the body: purging – in the era of bloodletting; draining – in the era of urban sanitation; emptying out – in the era of meditation. When New Age teachers tell us today that we have to get rid of the toxins in our body, it's a way of formulating this age-old problem: that there is too much inside us.

For the same reason, medical and spiritual procedures that promise to extract something from the body have always proved popular. Removing bacteria-harbouring teeth to restore 'mental health', or the gall bladder or appendix when there is no infection, are all stagings of a separation from ourselves. The immense appeal of exorcism films caters, in part, to the same taste for taking something out of us. We are all relieved when the evil spirit is forcibly removed from the body.

Gym visits may play a similar role, as the person runs or cycles for hours in order to avoid the feeling of physical agitation that would otherwise overwhelm them. When we speak of drives here, it is less to indicate some purely pleasurable activity than the craving of certain zones of our body for stimulation, which, when the stimulation fails, is experienced as uneasiness, tension and an unfocused sense of urgency: indeed, the very terms that many people might use to explain why they have to go to the gym.

Human beings inhabit a spectrum here, between the Grail and the Ring. At one level, our lives are defined by desire, by striving towards, by direction, all embodied in the image of the hand that reaches out. The old Grail story is just that, the idea of a quest for a single unique object that offers untold reward. It is no accident that the depiction of the Grail is so often accompanied by that of a hand stretching out towards it. But at another level, we are compelled not to acquire something but to divest ourselves of it. In *The Lord of the Rings*, the most successful children's narrative of our time, the key is not to get the ring but to let go of it.

Think also of what happens to those who actually manage to reach the Grail or the Ring or the elixir that they have been searching for. In countless stories, from the *Indiana Jones* saga to the *Pirates of the Caribbean* films, those who touch the prized object are destroyed, their bodies quite literally exploding or dissolving from the inside. The only route to salvation is to get rid of the object. This is the difference between desire – in which we reach out towards something beyond us – and drive – in which we try to attenuate an agitation within us.

Desire is in itself an achievement, a creation of the child, which situates our goals outside and beyond the body. But drive is more primary, and we could see desire, as Lacan suggested, as in a sense a defence against the drive, one that is not always so successful. The search for 'more' (desire) is at the same time a search for 'less' (drive). If desire emerges from the original experience of the drive, perhaps this explains why the hand is so caught up with both of them: as an image of desire, the hand reaches out, but as an organ of the drive, it is overbrimming with morbid excitation.

Hand technology legitimizes this ferocious regime of the drive. As our fingers tap, brush and scroll, we are not limited to gym time or the more obvious bodily practices labelled sport or exercise. We can do it anywhere, and, for many people, this 'can' is in fact a 'must'. When Pascal said that modern man is no longer capable of remaining at rest in a room, it is less because of an overabundance of information or distraction, than to spend some of this bodily excess that saturates us.

4

In the 1990s stoner movie *Idle Hands*, an evil force takes possession of those deemed to be society's 'laziest fuck-ups'. When the hero, who is obliged to tie up his hand to stop it from killing people, asks a friend for advice, he is told that 'The trick is to keep yourself busy . . . keep out of trouble. Idle hands are the devil's playground.'

This last phrase is of course a reference to the long tradition of warnings, both in religious and secular contexts, of the dangers of not keeping the hands occupied. In the Bible story, Adam and Eve are doomed to toil with their hands after eating from the forbidden tree. If work before the Fall was pleasant and effortless, it is now difficult and painful. Keeping the hands busy and labour are equated here, and until not so very long ago it was often argued that the less privileged should be kept at work on low wages for as long as possible, lest the demon of unrest were to take advantage of their leisure time.

These ideas obviously served the propertied elite who expounded them, and justified the long working hours and minimal pay that they enforced. But at the same time, and for many centuries previously, idle hands were seen as a threat both to the individual and to the social order. The hands had to be kept busy. Monks and nuns had to be occupied in all waking hours, less to perpetuate the fate of Adam than to maintain the 'knotting' of mind and body necessary for survival.

There is now a substantial literature on the history of what is called the 'civilizing' of the body. From religious advice to the courtesy books of the fifteenth and sixteenth centuries, people were told what to do and what not to do with their bodies, how to sit, how to stand, how to eat, how to converse and how to occupy the hands. Although such publications would be read only by those privileged enough to have access to them, they shed light on how the body was conceived and what was expected of it.

The sociologist Norbert Elias has described how gesture and bodily behaviour were progressively shaped by court society, with the effects spreading far wider to other areas of urban and rural populations. Courtesy books show the ubiquity of snot and saliva, as a nation of coldy people ate with their hands, generally taking food from the same bowls and the same joint, tearing off what suited them and using pocket knives to spear, wiping their mouths and noses on their clothes, licking their fingers, belching, farting, scratching, snorting and spitting. The sheer frequency of admonitions to avoid such activities is indicative of their presence, and Elias thought that over the sixteenth century things began to change here.

During this time, a new court aristocracy was formed with rigid social hierarchies. There was more emphasis now on not offending others, and the place of the body was altered. People seemed more conscious of bodily functions and their effects on others, and the demand for what was considered the right behaviour became more emphatic. Erasmus's *De civilitate*, with its instructions for good conduct and the discipline of the body, was indeed, after the Bible, the single most widely edited and published

43

book of the sixteenth century. The body was being increasingly regulated, although Elias was careful to stress that there is no degree zero here, and that human groups have always had symbolic codes that shaped bodily practice. It was just that these moved in a particular direction from the sixteenth century onwards.

Elias noted the gradual passage of the home from a unit of production to one of consumption: activities like weaving, spinning and slaughtering would over time be transferred to specialists and craftsmen. New chains of dependencies developed, which meant an attention to how one would be perceived. What were considered to be acceptable uses of the body would change simultaneously, and Elias pointed out how disgusted we are by someone eating noisily or with their hands, when there is no health risk to them, yet the thought of someone reading in bad light or being poisoned by gas – when the consequences for health are obvious – produces very different reactions.

Despite his many volumes on the civilizing process, it is remarkable how rarely Elias is cited by mainstream English-language historians and commentators. In fact, when he is referred to, it is almost invariably followed by a few condescending remarks: the work is too general, misguided, unruly, ill-fitting, a bit wild. Although there are clear instances where his historical data and his hydraulic model of emotion can be contested, this attitude seems to mirror quite uncannily the civilizing process itself that Elias spent his life trying to describe. Like the body he studied, his work must be contained, limited, corrected and taught manners.

The same language of containment and restraint is found in contemporary medicine. Gail Kern Paster argues that Renaissance

44

physiological theories echo and reflect this same passage of the 'civilizing process'. Depictions of the bodily vessels and spirits emphasize impulsivity and force, the energy that produces movement and feeling. Just as the limbs of the body required proper balance, so too did its interstices. An internal system of regulation is at play here, to civilize what Francis Bacon saw as a series of 'explosions' in the body and Erasmus as its physical impulsivity.

In many ways these understandings of the body moved from a model of permeability to one of enclosure. The body was frequently described as a porous and irrigated container, in which humors moved around while remaining open to the outside. Regulation of internal temperature, for example, depended partly on external temperature, just as the moisture in the body depended on the surrounding air via the processes of transpiration and evacuation. As the king's physician put it in 1615, our bodies are 'open to the air as that it may passe and repasse through them'. As Paster says, the body was experienced as a kind of moving sponge.

Although it might seem as if the days of humoral theory are long gone, it is still powerfully present both within medicine and without. Contemporary ideas about chemical imbalance recapitulate Renaissance notions of the balance of humors in the body, and in everyday life we speak of 'catching a cold', being 'filled' with some emotion or of being 'in good humour'. As Paster observes, these expressions are a direct legacy of humoral theory. They might also help us to understand why in countries where humoral theory was once dominant we so often talk about the weather: rather than seeing this as empty small talk designed to fill the potentially embarrassing moments when we encounter other people, it may once have transmitted information vital to understand and predict the internal state of the body. How else could we explain the fact that even today, the weather forecast not only tells us what tomorrow's weather will bring but describes, post hoc, what today's weather actually was.

As the humoral theories fell out of favour in the seventeenth century, the new physiologies that followed tended to separate the internal dynamics of the body from outside factors like the weather and temperature. The inside and the outside of the body were perhaps more rigidly differentiated, a process that scholars have explored in several different areas of culture, from medicine to philosophy to natural science and poetry. While it would be unwise to push this generalization too far, as the body's permeability has never been lost sight of, what we do witness is more and more advice to keep the hands away from the body, and to stop the fingers from penetrating its orifices. As one historian put it, hands had to be kept not only to oneself but also off oneself. Where it seems as if once it was not uncommon for the digits to enter mouth, nose and ears, this becomes progressively frowned on, and generated new reactions, such as disgust.

Changes in manners will generate new forms of aversion and what may appear repulsive at one moment may seem attractive at the next. Disgust is certainly governed by such symbolic processes, but we might wonder, with the psychologist Gordon Allport, why someone may have no difficulty swallowing their own saliva or sucking their own blood from a pinprick, yet would be absolutely unable to do this if the saliva was presented in a tumbler or the pinprick already dressed with a bandage.

—

The management of the hands was of course not simply something learnt from courtesy book or pulpit, and we have to guess from many different sources how they were kept occupied. Painting, drawing and sculpture give us some clues, but it is always difficult to disentangle from the poses struck by the sitter and the omnipresent iconography how the hands actually moved and what they held. Literature, together with civil and legal documents, is helpful here, as are the archives of contemporary fashion and jewellery.

Much of the material found by historians, however, is not directly relevant, as it focuses on hand use as communication: the handshake, the bow, the blessing and, more generally, the language of gesture. While this is fascinating in itself, it has less to offer in terms of clarifying how the hands fiddled and fidgeted, although, as we have already seen, there are several indications. In the medieval period, the regimenting of churchgoers' hands encouraged fixed positions, and visual media show hands resting on low belts or daggers. The frequent depiction of folded hands in front of the waist implies a certain restraint, as does the tucking of the thumbs into the belt.

The church is a useful place to start here, as it is a space where we find both the body and rules to be imposed upon it. Early worship involved a great deal of physical activity, with hands, arms, legs, eyes and voices all busy. The hands would often be lifted to heaven or held joined in front of the eyes, and groaning and sighing were common, perhaps rather like in some forms of present-day evangelism. Yet by the mid-fifteenth century, those in prayer kneel devoutly with hands held neatly in front of them, palm against palm. Deriving probably from Germanic practice, this new configuration had its roots less in religion than in the feudal rites of vassalage. When a vassal received a grant of land from his lord, he had to swear loyalty and fidelity by joining his hands together within those of his master.

47

The relation to the deity in church was thus imported from the privileged technique of feudal subjection. Although the person praying had only one set of hands, by implication these were contained within another invisible pair, both constraining and guiding. Like the supernumerary hand hypothesized by the alien hand patient, it did not have to be visible or tangible to have its effects. As the prayer custom spread, it came to represent man's complete dependence on God, and may also have been reinforced by contemporary Franciscan practice, where the priest's elevation of the host had to be done with joined hands.

This dependence was also, of course, a disarming, and just as prisoners would raise their hands to surrender, showing in the same gesture that they were not holding weapons, so the church-goers' hands would be inhibited from fiddling and scratching and tugging when they were in the joined position. Islamic prayer practice involved a comparable imperative: hand movements were choreographed quite formally, and during one phase of the prayer ritual hands were placed firmly and squarely on the ground, effectively annulling their field of action.

We can note here that the most dangerous parts of the body in church attendance were not just the hands but also the eyes. Just as David committed adultery with Bathsheba by watching her, so too the churchgoers' eyes might stray. How could their gaze be inhibited when it was directed lustfully at others? By the end of the Reformation, there was at least a protocol: hands held together and eyes closed during prayer. In Islam, the position of the eyes was also tightly prescribed: their focus was directed at the centre of the holy mosque in Mecca. Thus what had always been – and continues to be – a problem for the church was solved, in principle if not in practice.

While medieval art and literature show people doing things purposefully with their hands, it is during the sixteenth century that we find the expansion of hand technology as such. Rather than seeing hands beseeching or pointing or gesturing, we find more and more hands just holding things. There are now a lot of things in hands, from fans to gloves to pomanders and lockets. These are obviously expressive of social prestige and are saturated with coded meaning, but they also testify to the rise of a series of objects that will keep the hands busy and occupied.

The gloves that feature so frequently in portraiture from the sixteenth century onwards would transmit information through the values embedded in their material, their colour, their cut, their trimming and their length. But as well as these surface features, what mattered was how they were held, how they were put on, how they were carried, how they were taken off. In social interactions, the glove was eminently present, and its manipulation resembles in many ways that of the cigarette in the twentieth century.

Fans are also especially interesting here. Popular with both sexes, they were carried or worn suspended from belt or girdle, and were used at what seems to be almost all times when other people were close by: there had to be one in the hand, and they were endlessly manipulated and upgraded. Folding fans were widespread in the mid-seventeenth century, with faces, maps, mottoes and religious and political scenes printed or hand-painted on to them. By 1710 there were around 300 different fan makers in London and there was even a fan tax at mid-century.

Handkerchiefs were also carried in hand, and could reach immense sizes. Sixteenth- and seventeenth-century documents reveal a vast amount of activity around dropped handkerchiefs, as well as dropped fans and gloves, as if one of the functions of these objects was precisely to attach and then detach them from the body. The rites of gallantry and social exchange, especially between the sexes, could be mediated by these accretions: a glove might not be commented upon, but when it was dropped it could function as the point of contact. When Walter Raleigh supposedly laid down his cloak for Queen Elizabeth to avoid the muddy ground, we should understand this voluntary action as nonetheless a part of the broader practice of dropping.

As well as these accessories to be fiddled with and dropped, perhaps the most prominent hand object during this time was human hair. In Elizabeth I's reign, people might spend half a day at the barber's shop, and beards could be stiffened, perfumed, plaited, dyed, twisted and, later, double-pointed. Touching beard and whiskers may well have been one of the most common ways to occupy the hands, until well into the twentieth century. Dickens, indeed, could write of moustaches that 'Without them life would be a blank.'

———

The great popularity of gloves and fans was later enjoyed by snuff and snuffboxes, which were apparently fingered incessantly by those who could afford them. One wit satirized in *The Tatler* can answer questions in court only while snuffing, and composes himself by 'sometimes opening his box, sometimes shutting it, then viewing the picture on the lid, then the workmanship on the hinge'. When the court confiscates the snuffbox, he is 'immediately struck speechless, and carried off stone dead'. And Yorick in Sterne's *A Sentimental Journey* declares on receiving a horn snuffbox from a monk, 'I guard this box, as I would the instrumental parts of my religion, to help my mind on to something better.'

The idea of life without snuff and a snuffbox seemed unthinkable, and Richard Steele could write in *The Tatler* that 'However low and poor the taking of snuff argues a man to be in his own stock of thoughts, or means to employ his brains and his fingers...' The hand would feel automatically for the snuffbox secreted in coat or waistcoat. So where today we have mobile phones, there was once the fan, the hair, the hanky and the snuffbox, not to mention the cane and the umbrella of the next century, necessary, we are told, because 'the Englishman does not gesticulate when talking and in consequence has nothing to do with his hands'.

It is tempting to link this apparent ramification of hand objects with other changes in social structure. Historians from many different traditions have argued that a new set of interactions between people emerged between 1550 and 1800, as 'private life' was created, whereby meetings outside one's immediate circle would take place in the rapidly expanding network of coffeehouses, lodges, salons and societies. These new spaces required the visibility of personal objects that would signify status and rank, and as the trade in hand-held technology increased, so the desirability of the objects was transmitted to other areas of society. People who could never frequent a coffeehouse or a club might still be able to access cheaper versions of fans or snuffboxes, or take advantage of the growing trade in second-hand goods.

This affluence of the new technology was noted and ridiculed at the time. Everyone was busy with their glove, snuffbox, watch or pocketbook, and the information broadcast by possession of the object had to be nuanced by how it was used, how skilful, adept, clumsy or unaware the owner was. A tongue-in-cheek advertisement in *The Spectator* of 1712 offered classes on how to hold snuffboxes and take them out of the pocket in 'the most fashionable' way, as well as tuition on the different gestures appropriate to sharing tobacco with stranger, friend or mistress.

In more recent times, as snuff would graduate to the cigarette, all of these features would be patently present, and smoking would provide the hands with something to do at nearly all times, both public and private. The pipe both preceded and coexisted with snuff for many years, and would signify lower social rank, although its associations would shift during the nineteenth and early twentieth centuries. Unlike the ready-rolled cigarette, the paraphernalia and preparatory work that the pipe involved would be more time-consuming, but also more absorbent of the hands.

Like the fan or the glove, a cigarette could be held and manipulated in the hand, while functioning simultaneously as a conduit to social exchange and a means of punctuating time. Just as important was its pacifying function: beyond the circuits of nicotine arousal, having the packet or cigarette itself in hand furnished for many a strange sense of self-sufficiency. The world might be an unpredictable, frightening and unjust place, but I have what I need right here, just this little object, the possession of which does not depend on anyone else. In a stunning reversal of our earliest life situation, the smoker becomes the one who carries with him his own supply.

It is difficult not to notice that, historically, as we move beyond the cigarette in our increasingly tobacco-free world, something has miraculously taken its place: almost out of nowhere, mobile phones now colonize exactly the same social and bodily spaces. Some years ago, a man described a very awkward situation: he was smoking when his date leant forward to kiss him. As the kiss became more passionate, he couldn't bring himself to extinguish the cigarette, and felt an immoral oscillation between the wish to keep dragging and the wish to prolong the meeting of lips. If this double attachment could trouble him, how common it is today to find the same situation, but where the choice is not between partner and cigarette but between partner and mobile.

———

What about the pocket? When the psychoanalyst J. C. Flügel wrote his *Psychology of Clothes* in 1930, he noted that when wearing his overcoat he had no less than twenty pockets about him. Earlier pockets were not shaped to accommodate the hand, and in the eighteenth century it would often be held just inside the waistcoat, with the other hand rested on the hip. Women had expansive tie pockets, bags connected by a linen tape, worn in pairs under the skirts or apron, which were replaced by smaller integral pockets only much later at the end of the nineteenth century, and then by the celebrated external pocket known as the handbag.

Men's pockets held watches, notebooks, papers, pipes and tobacco, keys, knives, snuffboxes, hankies, pens, calling-card holders, napkin or toothpick holders, spectacle cases and small tools. Women's tie pockets could contain snuffboxes, scissors, notebooks, tweezers, combs, mirrors, hankies, keys, prayer books and other domestic objects. Although the topology of the pocket has invited interpretation as a symbol of both the vagina and the womb, it is perhaps worthwhile to also consider it as a device for restraining the hands. Rather than seeing the male or female pocket here as a conduit to bodily stimulation, it can have the exact opposite function of keeping the hands away from the body and from each other. It gives the hands a place to be.

This function of pockets becomes even clearer in those places where we cannot use them. On the tube, it is almost impossible to remain standing with hands in pockets while the train is in motion. And something peculiar now happens, observed by the sociologist Erving Goffman: on crowded New York subways, where women have had reason to accuse 'standees' of mashing, men may clasp the centre posts with both hands positively glued to them, high enough so that anyone would be able to see where they are. This strange and uncomfortable position exonerates the hands, as if to signify, 'Whatever happens, it wasn't *my hands* that did it.' Exiting the subway, hands can return to pockets, the internal rather than the external device of restraint.

When Adam Smith complained of the 'trinkets and baubles' so treasured by his contemporaries, he saw only futility in the new apparatus of enclosure. 'All their pockets,' he wrote, 'are stuffed with little conveniences'; and 'they contrive new pockets, unknown in the clothes of older people, in order to carry a great number'. Their value, he thought, was 'not worth the fatigue of bearing the burden', but he missed the fact that it was not fatigue here but merely a new form of necessity. The hands needed something to hold, to hold on to, to manipulate, and hence the simultaneous efflorescence of both 'convenience' and pocket.

The continual mockery of people's attachment to what is in their pockets throughout the eighteenth century cannot fail to echo our contemporary dependency on mobile phones, less as enablers of communication and information than as vessels and mediators of bodily tension. Despite the apparent mutuality of 'convenience' and pocket, hand in pocket is never hand in glove. It needs more, and hence the apparatus of keys, pens, coins and now phones that are jingled and fingered within them.

It is clear that many of these objects have lost their functional use: change may be kept without ever being used, just as the transcribing activity of pens and pencils may be increasingly diverted to mobiles. The same, of course, could be said of the fan or the umbrella. A fan would indicate status, could act as a conversation piece and might also transmit to friends or lovers the signals of the fan codes of the seventeenth and eighteenth centuries. But it would also fundamentally be something to hold, to manipulate with the hands, to interpose in social relations.

As for the umbrella, if we need it when it rains, we used to need it just as much when it didn't. Traditional umbrellas were often never unfurled, functioning rather as part of the hand technology of the late nineteenth and early twentieth century. When Neville Chamberlain flew to Munich and then travelled by rail to Berchtesgaden in September 1938 in his effort to attempt a peace pact with Hitler, he carried with him a neatly rolled umbrella at all times, although it was obvious he would never have to actually use it in the plane, the train or the car that met him at every other stage of his journey.

—

As we track the discourse on hands over time, we see a slow but significant change. If keeping the hands busy was deemed imperative in many eighteenth- and early-nineteenth-century sources, it was to avert danger, sin and sloth. In Isaac Watts's famous verse, penned in 1715:

In works of labour or of skill
I would be busy too;
For Satan finds some mischief still
For idle hands to do.

Idleness itself was now created as a new concept, with devotional literature linking it to vice, and lust appearing, as one author put it, 'at those emptinesses where the soul is unemployed and the body is at ease'. Such associations are perhaps unsurprising in a period where there was more and more reflection on the meaning of labour and occupation. The so-called Protestant work ethic set a premium on time-consuming and steady work, and so anything that escaped or exceeded this would require identification and proscription. The enemy had to be given a name.

Historians have argued that the distinction between work time and leisure time owed much to the Factory Act legislation of the 1830s and 40s, which contributed, in part, to the establishment of more regular working hours. Where leisure could once be a privilege of the elite, it now came to signify a more common space, both agreeable and hazardous, containing a spectrum that extended from self-improvement to debauchery. It is perhaps significant that it is during this period that we witness a new ramification of the expression 'the hands' to indicate workers. As hands had to be kept busy through work, the worker, in turn, could be reduced to a pair of hands. For a worker to be anything more than this was potentially dangerous.

But as we move from the later nineteenth to the twentieth century, occupying the hands becomes linked less to avoiding danger than to pursuing health for its own sake. Soldiers returning from the Crimean War and then the First World War were encouraged to knit and embroider, with the idea that a constant repetitive activity would have calming effects on the nerves. If Johnson could write in *The Rambler* that he liked to see 'a knot of misses busy at their needles', because such work provides 'a security against the most dangerous ensnarers of the soul, by enabling themselves to exclude idleness from their solitary moments', an 1888 edition of *Dorcas Magazine* specifies that 'The quiet, even, regular motion of the needles quiets the nerves and tranquillizes the mind, and lets thought flow free.'

This would soon become incorporated into the many forms of occupational therapy, where rhythmic, often monotonous activity was encouraged as a route to rehabilitation and recovery. The emphasis was not only on the temporal aspect of craft, but also progressively on the emotional and spiritual significance of the created object. It is interesting that the spread of occupational therapy coincides with the popularization of images of factory work, where hundreds of people are depicted engaged in mind-numbing, repetitive tasks, as if the very factors which, for many, dehumanized work would now become seen as positive and intrinsically therapeutic.

Students were sometimes encouraged to knit if there was any pause during classwork, and early-twentieth-century doctors could vaunt the ticking noise of steel knitting needles for calming the nerves via a hypnotic effect. Knitting was 'quieting' and 'nerve-smoothing'. As one knitting boy wrote in 1918, 'The boys in our room that used to sit and fumble their ink-wells, or tap their pencils, or tinker with their rulers, or maybe flip bits of art-gum at you when someone was reciting, are so busy with their knitting that they never fidget or misbehave.' Knitting could of course come to have any kind of personal meaning and significance for the knitter, just as it could take on a variety of different connotations culturally. Used in some periods to construct gender differences, it could be linked to speech and silence, to industry and to poverty, to war, recycling and charity.

Reading through accounts of knitting practice, it is striking to see how frequently it is associated in a series to other activities – knitting and talking, knitting and watching, knitting and listening – as if it is somehow necessary in these situations to keep the hands occupied. Even today, guests may bring knitting to a social gathering, and many nineteenth-century sources are quite explicit about this as a requirement for social engagement. In the film *Idle Hands*, after the protagonist receives the advice to keep his possessed hand busy, we see him sitting on a sofa between his stoner friends, knitting needles firmly in hand.

Although the accent has moved from danger to health, knitting still always retains its place as a hand technology and hence its value as a defence. To knit is to make, to create, to share, to participate, yes, but it is also to ward off, to block, to keep in check and, perhaps, fundamentally, to bind. 'My mother hauled me to a knitting store to find something to keep my fingers out of my mouth,' says one lifelong knitter, and it is well known that the practice can be adopted in order to keep the hands 'out of food or off cigarettes'. Or, to put it another way, to stop them from returning to the body.

5

When friends and acquaintances asked me what I was working on during the preparation of this book, my reply that it was to be an essay about hands produced the almost invariable response, 'Oh! A book about masturbation!' The association appeared so intractable that it seemed foolish not to at least devote a chapter to this subject, although I was tempted to just cite the well-known vignette about Bruno Bettelheim. Lecturing to a crowded hall, he noticed a student knitting and informed the class that her manual activity was a substitute for masturbation. 'Dr Bettelheim,' she replied, 'when I knit, I knit; when I masturbate, I masturbate.'

It is true that the thousands of admonitions over the last few centuries to avoid idle hands can be interpreted as proscriptions against masturbation. In some places, the link is made explicit, with schoolboys banned from keeping their hands in their pockets lest they stimulate the genitals. If we believe the historian Thomas Laqueur, it was only really from the early eighteenth century that masturbation became big business, a sin that could lead to death, disease and blindness. Prior to that, it was deemed injurious every now and then, yet without the massive moral approbation it would receive in later times.

With the influence of psychoanalysis on paediatrics in the early twentieth century, masturbation would lose many of these connotations, yet even in the 1940s and 50s, children's hands could still be bound with string or stuffed into aluminium mittens to prevent their return to the body. The violence of this action is brought home by Solzhenitsyn in *The First Circle*, where he describes how prisoners were forced to sleep with their hands away from their bodies: 'It was a diabolical rule. It is a natural, deep-rooted, unnoticed human habit to hide one's hands while asleep, to hold them against one's body.'

It is important to distinguish here the disparate forms that masturbation can take. Stimulation of the genitals is preceded by a variety of practices of bodily touch, and the place that such activities occupy can be radically different. Picking at a bit of skin or pulling out hairs might not elicit any of the feelings of worthlessness or despair that often accompany genital masturbation. Likewise, in our era of 24/7 nappies, how many babies reared in the developed world actually manage to reach their genitals with their hands? The discovery of the sexual organs and what they can do may come later, and when it does, these shocking revelations will preoccupy almost every child.

Masturbation is also widely misunderstood to be an access to pleasure. The genitals are aroused to produce a nice feeling. This might be true, but in so many cases we find something quite different. When the mother of a little girl admonished her for pulling up her knickers so that they stretched into her labia, she commented, years later, that doing this 'was not to produce a feeling but to get rid of a feeling'. The brief pleasure sensations of genital masturbation can function as a barrier to other more powerful and overwhelming forms of bodily agitation and anxiety. People tend, after all, to masturbate not just when they are aroused but when they are anxious, just as arousal itself often follows experiences of loss and pain.

In their early studies of infant feeding, researchers at Yale in the 1930s were surprised to see how often the male babies got erections during the process. But to their greater surprise, they found that these never occurred during states of pleasure such as satiety or play with the nipple. It was only when the feeding was delayed or difficult or interrupted that the penis became engorged. The physical signs of arousal appeared in these situations of frustration, rather than in those of satisfaction or fulfilment.

In the modern world it is well known that the most common use of the Internet is to view pornography. But what precedes this appeal is never indifference. When we explore in analytic work the specifics of why a patient browses porn at the precise moments that they do, we always find an initial lowering of mood, an office conflict, a sadness that may be held away from consciousness. The fact that it is the Internet that allows this is also not accidental, as it is so often the Internet that is the very conduit of the initial anxiety. Bombarded with incessant emails, always open to the demands of other people, pornography can function as a separation, a way to attenuate the feelings of invasion and pressure that Web use introduces.

Masturbation in this sense is a way out, an escape mechanism, and its analgesic quality may be used to counter any form of physical or mental unrest. When the experience of sexuality proves unmanageable and overwhelming, genital stimulation can be a way to try to localize it and hence contain it. The psychoanalyst Karin Stephen pointed out many years ago that the only force strong enough to fight against sexuality is sexuality itself, with the suggestion that techniques of renunciation could become sexualized in themselves. But non-renunciation can also have this function. As another patient put it, 'I masturbate to get rid of sexual feelings.'

—

If infancy always involves the hands touching, scratching and rubbing parts of the body, the transition to the genitals is not a smooth one. It is not a question of simply finding a new site, as the sexual organs and how they function have to be discovered. Ejaculation, discharge and orgasm are staggeringly traumatic events, and a whole apparatus of our culture is there to try to frame and encode them.

A few years ago, an online discussion group made up mostly of psychoanalysts debated how best to normalize the experience of early sexual arousal. If you return home to find your child masturbating, they asked, what should you say so that there are no traumatic after-effects? How can the child be made to feel good about what he or she is doing? The reiterated suggestion that they should be told that it was all quite normal and that everyone does it totally missed the point: after all, it wasn't just that the child was masturbating, but that the parent had found them doing so. To remind them of this inter-subjective dimension, the writer Hanif Kureishi interjected that the real question ought to be: if your child returns home to find you masturbating, what should you say to be less trau-matized yourself?

The effect of sexual awakening is, of course, by its very nature traumatic, and to claim otherwise is fanciful. Nothing can pre-pare the child for the new sensations and engorgements in the body, and each child will have to find ways to process and make sense of these experiences. For some, this will entail a movement away from the genitals, so that other parts of the body suddenly become the child's focus.

Many boys have the curious thought that one day they might wake up with 'big hands'. The comic potential of this is obvious, and it certainly possesses a certain historical pedigree, from the massive hands of the actors in medieval editions of Terence's plays, to Leonardo's painting of the lady with ferret and the Parmigianino self-portrait, where the sitter inflates the size of his own artistic hand. Trauma here becomes mitigated, as the idea of a body part getting larger during the night migrates from penis to hand.

But the hand can retain all the opacity and shock linked to the genital metamorphosis. In one of the most celebrated scenes of transformation, the young protagonist of *American Werewolf in London* shrieks in pain as he watches his hand elongate before him. In the Spiderman saga, the adolescent Peter Parker stares at his hand with both wonder and horror as it shoots out a gooey substance. The story deals with bodily change and how a teenager can come to terms with it, just as so many of the superhero narratives revolve around the question of trying to process new sensations in the body, usually glossed as 'powers'.

Where boys might dream of waking with big hands, girls sometimes imagine, on the contrary, not possessing an enlarged hand but being carried around inside one. The image recurs with some frequency in daydreams and in culture, and we could think of Fay Wray or Naomi Watts both kept and kept safe in King Kong's palm. The vast hand holds them prisoner while also protecting them from danger, a confluence that we find in other areas of female fantasy life. The motif here is linked to the classic figure of the woman who can tame the beast: while everyone else wants to kill or capture it, she alone knows that it has a good heart and only poses a threat when provoked.

This unspoken knowledge shared between her and the monster could be understood as an Oedipal love: the daughter is the only one to fully understand the father, and recognize that beneath his apparent power there is nothing more than weakness and lack. But at another level, the enormity of this creature is perhaps an incarnation not of the father but of her own sexuality. Whereas boys may be unsettled by the physical changes in their body, fear of experiencing desire is much rarer here than it is in girls, who may be terrified at their own erotic sensations. Kong's hand may be so huge because it is an image of exactly this arousal, which exceeds ready quantification.

For other children, there is less a displacement away from the genitals than a direct engagement, one that may allow a certain separation from the family, as a private space is now constructed around the new activity. The child may spend more and more time by themselves, and establish new standards of domestic privacy. Doors are locked, 'Keep Out' notices posted and time alone defended.

In his history of masturbation, Thomas Laqueur is perplexed by the problem of why it came to represent such a danger in the eighteenth century. He notices a correlation between the new private spaces established to allow silent reading and the threat now posed by bodily stimulation. Prints and drawings of masturbatory practice from this period show with uncanny frequency the presence of the printed page, with a book either in the person's hands or at their side. The creation of private space and the 'danger' of masturbation may thus share a certain historical root.

Studies of early masturbatory behaviour have made an observation here that echoes this idea. Researchers like Eleanor Galenson, Herman Roiphe and René Spitz found that, contrary to expectation, masturbatory practice did not correlate with fusional behaviour with the mother but with its opposite: once the child was engaged in genital masturbation, they seemed better able to withdraw from their dependency. Masturbation and autonomy were associated rather than opposed. The child was enjoying rather than being enjoyed, finding alternatives to being the object of parental pleasure.

At the same time, new creative capacities can emerge, as a way of processing the bodily sensations. We could think here of all the games which involve building a secret space or den, a place prohibited to adults, or the idea of suddenly finding a secret treasure hoard. When we read that Sleeping Beauty on the eve of her sixteenth birthday found a funny little room hidden deep in the castle, it would be prudish not to interpret this as a reference to her own secret room, and the blood that subsequently flows when she touches the spindle as a reference to her own hymenal or menstrual blood.

So many fairy tales feature an opening that is revealed where no opening was known to exist. The entrance to Aladdin's cave appears out of nowhere, and the sliding rock that grants access to Ali Baba's den couldn't be found until the moment the 'Open Sesame' formula is repeated. More recently, in *The Hobbit*, Bilbo and his friends scrutinize a cliff face for days without finding anything, until a play of light allows the discovery of the door's outline. Beyond it lies the gold they dreamt of.

The boyish search for a secret hoard of hidden treasure, still pursued by many an adult with their metal detectors, evokes both the maternal gold guarded jealously by the father, but also the sudden rush of 'gold' that occurs with the first ejaculation. How could such a treasure have been hidden right there under their noses! Teenagers and adults still often cheer and exclaim out loud in cinemas when, during so many action films like *The Matrix* or the *Terminator* series, the hero uncovers a cache of automatic weapons in some hidden place.

—

That the discovery of the treasure hoard takes place by chance is, as Selma Fraiberg pointed out, exactly what usually happens with the first orgasm. It just happens, and this accidental, contingent quality is found in many fairy tales. Aladdin rubs the lamp, and a genie magically appears who will grant all his wishes. An act of accidental rubbing thus has major effects, and we could note how the genie is invariably depicted in the act of engorgement, swelling up out of the tiny lamp.

The secret doors, rooms and cavities that appear in these tales are geographies of the body, and they chart the child's explorations and discoveries. The magic that they release is dangerously good, and the heroes and heroines often have to pay a price for their enjoyment. Fraiberg observes that whereas in the older folk and fairy tales, this magic is released by use of the hands, in later stories it is through an act of deciphering. A map has to be interpreted, as in *Treasure Island*, or a code, as in *National Treasure*. In both cases, once the text is unencrypted, the place where something is hidden will be revealed. The Tintin stories are absolutely filled with such devices, and there is always a code or map that needs to be deciphered.

But why the motif of deciphering? Why would a code have to be cracked to access some bounty? Does this really resonate with bodily experience? We could remember the story of Oedipus here, who before marrying his mother has to answer the riddle of the Sphinx. There is thus an enigma prior to what might have seemed like a reward.

One way to understand these maps, codes and riddles is simply as signs that something has to be made sense of. In our relation to traumatic sexuality, we never have the right symbolic tools, and so we have to crack some code, find some meaning where we are confronted with an impasse. There is both the presence of a bodily sensation and the absence of a meaning. Through the invocation of ciphers and encryption, we are symbolizing the problem of symbolization itself.

In many of these stories, there is luckily someone to help the hero or heroine. Difficult as their quest or mission might be, they will invariably encounter a donor, often an elderly figure, who will give them some magical objects. We could think here of Perseus, who receives his bag, sword and magical cloak from the Hesperides, and also, more recently, of Q in the James Bond films, who appears in each instalment to give 007 some new life-saving gadget.

Fairy tales – ancient and modern – fuse two different threads here. First of all, the question of the body and its mysteries. Second, the Oedipal relations within the family. If the geography and the magic in these stories are about the body, the dynamics of gift giving, competition and outwitting circumscribe the turbulence and drama of desire within the family. The donor figure, like Q, is the external agent who offers mediation, yet we can note that in so many versions of this narrative the weapons and gadgets have to be taken by force or trickery, as was the case for Perseus.

Just as the child has to take something owned by the father to be able to satisfy the mother, and to be armed for the future, so in these tales, as Fraiberg saw, there is always someone with greater powers than the boy. Whether the child's weapons are bestowed – as for James Bond – or coerced – as for Perseus – there is always a transfer, a handing over, of power. Aladdin and Ali Baba steal someone else's secret to gain access to their riches. And as Aladdin says, these riches have only one purpose: to make his mother happy.

⎯

Scholars like Margaret Schlauch and Marina Warner have traced the revisions and rewritings of these tales, which repress the original incestuous motifs. The Cinderella story, for example, is part of a long tradition of narratives in which a father is told by his dying wife to remarry only a woman as fair as she is. At some point he realizes that this person is the daughter, and insists that they wed. This incestuous union is averted only by the daughter cutting off her own hands, and sending them to her father.

In these stories the requirement that the king wed only one as fair as the queen also takes the form of the demand that the future wife is the one whom her ring or clothes will fit. The king is unable to find such a person until the daughter is older, and then the love starts to gather its transgressive momentum. As the daughter then refuses this mandate – which, we should note, is a maternal and not a paternal one – she makes the sacrifice of her hands.

In the Italian story *La Madre Oliva*, the dying queen exhorts her daughter to continue her work of almsgiving. One day the king sees her and proclaims, 'My daughter, I am in love.' 'With whom father?' 'With your beautiful hands.' She promptly cuts them off, sending them to him in a gold vessel. In the medieval romance *El Victorial*, as father and daughter mourn together, he tells her he should die if he did not have her at his side, given her resemblance to her mother. Later his love becomes more urgent, and he wishes to marry her. When he kisses her hands, she commands a trusty servant to cut them off. In the German *Herzog Herpin*, the daughter says, after cutting off her hands, 'My mother had all her limbs, but I have none.'

This motif has been understood in different ways: as a mutilation to make herself unlike the mother and hence no longer subject to the mandate, or simply as a way to make herself unfit for marriage. Hermione Thompson suggests that it could also signify a rejection of the very dimension of agency: if this is what the hands embody, severing them shows how agency is precisely what the daughter has lost in the prison of her parental imperatives. Given the father's emphasis on the beauty of her hands in so many of the stories, we could also read her act as a kind of psychoanalytic interpretation of his desire: 'If this is what you love so much, here, take it'; as if to show him that his desire is aimed not at her person but only at a part of her body.

The daughter here is, after all, initially situated in the void left by the absent mother, so the gesture sends back the message that the father loves something 'in her more than her'. The priority of the mother is a feature that recent versions of the Cinderella story have recognized. In the latest Disney film, we see the mother singing a lullaby to the young Cinderella before she dies. Later, at the close of the film, the prince searches for the girl whose foot will fit the slipper, and it seems as if all hope is lost as everyone who has tried it on has not been 'the one'. Then he hears a voice singing from a tower. It is of course Cinderella, but she is singing the lullaby that had functioned as the acoustic cord between herself and her lost mother.

This detail shows that the love between her and the prince is built upon the love between her and her mother, rather than the foppish figure of her father. In a similar vein, Disney's retelling of the Sleeping Beauty story *Maleficent* operates a comparable shift. We witness the troubled witch's participation in the girl's childhood, and after Aurora falls into her long sleep, the kiss that wakes her is not the prince's but Maleficent's. In both versions, the priority of the mother has been reassigned.

In the Cinderella story, the slipper has been interpreted as a representation of her sex, and a pornographic film has apparently been made which depicts the obvious consequences in terms of the prince's search. But what we find here is perhaps more of a trace of the original narrative, with the slipper that she leaves behind at the ball taking the place of the hand that the daughter leaves for her father. As the story is retold, the refusal of a marriage is transformed into the very conditions of a marriage, and the incestuous motif is lost.

In this sense, the abandoned slipper is like all the other objects that characters in fairy tales drop behind them to delay their enemy. Whether it's a comb or a bracelet or a tooth, it is thrown not to entice or intrigue the pursuer but to slow them down, to give the heroine those precious extra moments to escape.

It is curious that the slipper is also the principle of so many games played by children, in which an object is deliberately lost and then found, or detached and then reattached. We could think of the practice of skipping, or even of the 'Fort/Da' game we discussed earlier. Something leaves the body and then finds its way back to it. In the severed hand stories, they tend not to, but once we recognize the initial situation of the father's love for the dead mother, we might wonder whether the hands that the daughter sends to him are not, at some level, also those of the mother.

—

We have seen how our own bodies can become colonized by the bodies of others. Children rock themselves as they have been rocked, stroke themselves as they have been stroked. Autoerotic activities might appear to be self-absorbed acts of pleasure, but they can equally be exercises of cruelty and revenge, as one's own body is treated as if it were the body of someone else.

An eight-month-old girl described by Eleanor Galenson would fiercely bite her mother's hand or body on minor frustrations or for no apparent reason. The mother then started to take her daughter's hand and place it in the latter's mouth, as if to signify, 'Don't bite me, bite yourself' and 'When you bite me, I get so angry I would like to bite you.' Within three weeks the biting attacks stopped. Now, when angry, she would bite her own hand, and, some time after this, pull at her own hair.

Masturbation here can be a vehicle of vengeance, and a patient explained how when touching herself she focused her thoughts exclusively on her hatred of her mother. Another man described his feelings after masturbating with the words: 'It's like having killed someone.' This may shed light on the frequent self-reproaches that emerge in its wake: 'I should just die,' repeated another patient, and the person may experience the most intense sense of worthlessness, degradation and ruin.

Freud emphasized the difference between masturbation as a physical pursuit and the fantasies that accompanied it and often made it possible. He argued that the sense of guilt and worthlessness that might follow were linked to the fantasies and their hidden incestuous content. One of my patients would delay ejaculation by summoning up the image of the Union Jack flag, and afterwards feel despair and dejection. Exploring this peculiar practice, it became clear that it was linked to his having overheard his mother describing her experience of sex: 'Well,' she had said, 'one just has to lie back and think of England.'

To avoid any conscious connection between a parent and sexual activity, a robust censorship is required, and this can entail the eclipse of an entire figure in the fantasized sexual scenario. For my patient, indeed, the British flag replaced any image of a flesh-and-blood body, and in many cases, Freud observed, instead of picturing some paternal figure touching her genitals, a girl might conjure up nothing more than the image of a hand.

Michael Balint discussed the case of a patient who would masturbate to the image of just this: a stranger's hand. She herself would take no part in the scenario, but remain completely passive, as she had to if her secret was not to be discovered. He notes how common it is that people will set a date when they are to give up masturbation, with the idea that a new era will begin, only to then see their resolve turn into postponement. Another of his patients made a diary with private hieroglyphics, including signs for coitus, masturbation and wet dreams, and we could add that biographers today are forever turning up such idiosyncratic lexicons.

The whole idea of a new leaf perhaps has its origin in this struggle to go beyond masturbation, and Balint catalogues the many practices that are used to distance the act itself. At first, the hands may be suppressed, and in their place come fidgeting with the feet, rolling on the stomach, pressing the thighs together, squatting on one's heels, or crossing the legs. Keys or coins can be jingled in one's pockets, pencils chewed, nails bitten, noses picked, and lips and mouths variously manipulated.

Although such activities may serve as diversions from genital masturbation in some cases, Balint was perhaps too quick to collapse them all together. A meaning later attached to them might have been initially absent. A boy may pick his nose with impunity as a child and then, after he starts genital masturbation, experience acute shame if he is found nose picking. The action is technically the same, but a new meaning has been added to it.

Any discussion of hands needs to factor in what we learn from masturbation, as it impeaches many of the grand claims made about the human sense of touch. Philosophers have often elevated the hand to almost divine status due to what is taken to be its reflexive capacity. When we touch, we simultaneously feel an object and feel our hand feeling. The dense concentration of receptors in the hands encourages this, and as we feel ourselves feeling, we apparently enter a new stratum of consciousness, especially when hand touches hand.

But how many people can feel their hand feeling their genitals? One study, indeed, referred to the 'sensory effacement of the hand' in masturbation, and the tendency of the hand to delete its own feelings. What we find here is in fact something that could be taken as paradigmatic of human sensory experience. The term 'extinction' is used in neurology today primarily to refer to the sensory depletion following brain injury, yet in the 1940s and 50s researchers found it to be present in both those with neurological damage and those without.

Where most traditional studies of sensation had started with the application of a stimulus to parts of the body and then recorded the results, Morris Bender and his colleagues worked with double or triple stimuli, so that the person would be touched, pricked or poked in more than one place at once. A stimulus would be applied to the face, for example, while another was applied to the hand, and it became clear that simultaneous stimuli function to mask and negate each other. Comparing stimuli to the thigh, nose, face, breast, penis, foot, buttocks and ear, Bender found that the least responsive part of the body was in fact the hand. More than half his adult subjects failed to register any stimulus at all to the hand, with even higher rates for children, even when they were told that the stimulus would be applied. The face proved out of all areas to be the most dominant region.

Comparing the experiments with brain-damaged patients and controls, the results showed that extinction was present in both groups. Bender saw this as a refutation of the popular theory of the nervous system as a mosaic of neatly specific centres and junctions connected by discrete pathways. On the contrary, his experiments suggested a system in which signals are perpetually affecting each other, suppressing, negating or rerouting sensory perception. Although we might only notice extinction in those with brain injury, it is present, he concluded, in every sensory process.

The hand has not fared well here, yet the balance of pain introduced by the stimuli in these studies illuminates both the question of masturbation and that of manual extinction. As my patient pointed out, she was using feelings to get rid of feelings. Pickpockets operate with exactly this logic: they will stimulate one side of their target's body with a push or a bump while they simultaneously stimulate the other side, extracting wallet or valuables. With the hand, varying its sensitivity can serve to focus sensations elsewhere or, at times, to concentrate them within itself.

When spy Harry Palmer is being forcibly brainwashed in *The Ipcress File*, he is strapped to a chair and assaulted with a barrage of invasive sounds and images. His sensory fields are defenceless, until he manages to pressure a small nail hidden within his hand into the flesh. Although his eyes still see what they see and his ears still hear what they hear, the pain focused in his hand allows him to be somewhere else. It is this simultaneous stimulation that saves him, this tiny point of self-generated pain.

Don't we see something similar in the way that children so often confuse the verbs 'scratch' and 'itch'? In terms of the receptors in our skin, an itch and a scratch are entirely different things, involving different pathways, and there is no good physiological reason why a scratch should relieve an itch. Yet this never stops people from scratching, and from calling this scratch an itch. Isn't the same principle at work here as in Palmer's strategy? If for the spy the pain generated by the nail was located in a different sensory field to that caused by the sounds and images, here both stimuli converge at the same part of the body surface. One temporarily negates the other.

We might feel some relief after scratching, but then, as with other kinds of self-stimulation, there may be regret and a sense of failure. The hands, once again, have been doing things they are not supposed to do.

6

In the vast literature on the human hand, the question of tools invariably arises. Tool use, we read, not only differentiates us from beasts – who can use tools but generally neither store nor share them – but also pushed the brain forward in its path towards evolutionary success. When we stopped swinging from trees, the hands were now free to make things which would, for Darwin, allow a new kind of mastery. But rather than the brain propelling these changes, it was the changes themselves which would retroactively affect the brain. As Stephen Jay Gould put it, the palaeontologists searching for bigger and better skulls were looking in the wrong place: they should have been looking at the hands.

The new upright posture freed the hands from locomotion, and allowed them to start manipulating things in apparently unprecedented ways. While recognizing the significance of this shift, we could nuance the argument: rather than seeing tools as what allowed the hand to change, driven by some sort of evolutionary force, the advent of tools can be understood as a way to keep the hands busy, to ground and focus the excess in the body. We invented tools to do all sorts of things, of course, but also fundamentally to give the hands something to do.

In the famous scene at the start of *2001: A Space Odyssey*, when the chimp discovers tool use for the first time, the bone it wields is immediately employed to kill another chimp. If the excess we have described is linked to sexuality, it is also caught up with the question of violence. And we can remember here that if we think of the hand itself as a tool, this lethal dimension is never far away. The hand that generously gives or thankfully receives is also the hand that can strike, punch and pound. Elementary weapons such as sticks and stones simply increase and extend their power. As Elaine Scarry points out in her remarkable book *The Body in Pain*, popular culture often makes tool and weapon indistinguishable. The hand that pounds a face and the hand that pounds dough are fundamentally the same hand, just as is the knife that enters the cow and the knife that cuts the meat.

A clenched fist is both tool and weapon at once. Scarry notes that if the first artefact was found to be a bowl rather than a hammer, this would not change the fact that the first artefact was a tool, since in order to make the bowl the hand had to be first made into a tool: in other words, used as a shaping agent. What subsequently distinguishes tool and weapon is simply the surface on which they fall. If this surface is sentient, we would speak of a weapon; if non-sentient, then of a tool. As Scarry observes, we see great progress when the hand holding the knife above Isaac in the biblical story then moves to a goat, but it is still a weapon until we substitute a block of wood for the goat. The aim is now less an act of wounding than of creating. The knife is transformed from a weapon into a tool.

The distinction here is complicated by the question of representation. The patient in Gestalt therapy who is encouraged to express their anger at a parent by beating a cushion can hardly be said to be using their fist as a tool rather than as a weapon, even if the fabric surface is non-sentient. What matters is what the cushion is supposed to represent, and, in equal measure, what the beating itself is supposed to represent. Or, to take another example, when a frustrated worker returns home and forcefully throws an empty bottle into the dustbin, is the hand tool or weapon? The question of intention here is unhelpful, as they might just explain the action as instrumental: they wanted to get the bottle into the bin, end of story.

—

If sticks and stones preserve a link to the hand, don't later weapons like bows, arrows and firearms make the distinction of tool and weapon less ambiguous, and change the role of the hand itself? Archery relies on the coordination of hand and eye, and evokes the visual grasping that we discussed earlier. Its distinctive action of opening the fingers, of letting go, is also perhaps linked to the question of release, of the ability to separate from something. But it is the firearm that brings us back, paradoxically, closest to the hand.

When teenagers and young men cock their hands as guns as they replay the famous 'You talkin' to me?' scene from *Taxi Driver*, it is difficult not to associate the configuration of the hand with the gun itself. The word 'firearm', indeed, includes 'arm', and the barrel finger is the same finger that points and blames. Gun use itself may be facilitated by this association. If the weapon is assimilated to the hand, killing may become easier. Anyone interested in violence cannot ignore the fact that many of the world's most popular computer games involve eliminating other people, moving an icon to either point at or touch the adversary.

In the weeks that followed the 2015 *Charlie Hebdo* killings in France, when two gunmen burst into the offices of the satirical magazine and murdered twelve people, many of my patients talked about the events with sadness and shock. Yet by far the most frequent response was a mental replaying of the events, as if from a computer console, where the movement of gun from person to person was endlessly repeated. There was a fascination with the logistics of the attack: how could so many people be shot in such a short time? How did the gunmen rotate their weapons? Did they preconceive an order of victims?

In all of these fantasies, the point of view was not that of the victims but of the gunmen. The replay was from their perspective, exactly as in a computer game one has to shoot multiple targets in a very limited time, moving from one to another as fast as possible. Note that these fantasies were described by the very same people who felt sadness and shock, and it reminded me of the way that *Schindler's List* had generated similar reactions. When this film first came out, it was less the plight of the concentration camp victims that transfixed people than the camp commander's random shooting of them from his watchtower. It was his enjoyment of this horrific target practice that many of my patients at the time identified with.

As with the *Charlie* fantasies, replaying the act of shooting was always dispassionate. There was no welling of emotion, no regret, no pity for the victim, just a cool, methodical removal of the target. Computer games rely on the same principle, and the most feted male heroes of our times are those who can take out the opposition without any expression of pain or remorse. The aim is to kill as many people as possible in the shortest time possible. It is worth recognizing that this is the aspiration of the millions of people who game for several hours each day. Although contemporary culture puts a premium on emotional expression, deleting it can become in fact a much more powerful source of satisfaction.

Anyone who travels on public transport knows that there are always those who are waiting for the slightest opportunity to explode into aggression. If you sit or stand too close, if your bag brushes theirs, a torrent of abuse or physical violence may follow. If shootings are extremely rare, gun use is nonetheless present in a peculiar way. When Harmony Korine's film *Spring Breakers* came out a few years ago, a patient described his amazement that the filmmaker had understood his daily experience. The film soundtrack is regularly punctuated by the sound of a gun's closed-bolt reloading mechanism, regardless of the actual events onscreen. This was my patient's personal soundtrack, walking down the street, shopping in the supermarket, travelling on the train: a gun constantly being reloaded, and sometimes accompanied by the idea of shooting those around him.

It turns out that a lot of people go through the day with similar sounds in their head: a kind of gun soundtrack, with a focus on the loading rather than the shot. Violence is not externalized but confined to the private auditorium. The sounds become accentuated on entering a public place, as if to index the dangers of proximity with others. As another patient described it, 'When I'm out there, there are only two choices, to defend or to attack.'

—

That society depends on the potentiality of violence is indisputable. Both the history of human conflict and the imagined utopias of the future illustrate this clearly. Almost every single science-fiction narrative about the future depicts an 'ideal' world predicated on a horrific secret: unwanted citizens are eaten, killed, recycled, discarded; memories and emotions are erased or altered. It's always the same structure: perfection or happiness is rendered possible by some voluntary or involuntary act of sacrifice or loss.

Sometimes the secret is guarded by elders, sometimes it is known by the whole group but not spoken. Whatever the case, there is always an 'other side' to things, to social order, and maintaining appearances comes at a high price. Protagonists in these stories are often confronted with a choice: to let life continue as it is or to disrupt it, at great risk to themselves and others. From *The Hunger Games* to *The Cabin in the Woods*, from *The Purge* to *Rollerball*, the peace is kept by one communal act of violence.

Beyond the realm of science fiction, it is remarkable how almost every single adult film today – excepting genres such as romcom – revolves around the question: when is an act of violence justified? Finding counter-examples is more manageable than finding examples, since they are just so ubiquitous. At what point can a man or a woman kill? At what point is revenge acceptable? At what point should one move beyond forgiveness? When is it permissible to defend one's family? Historically, one could map a thread from the old gunslinger's question of when he should take up his gun again to the contemporary drone operator's question of when he should press the fire button – or refuse to.

In America, these questions are inflected by the issue of the president's safety. There are literally hundreds of films, TV series and books where the life of the US president is in danger, from a sniper, an internal or external plot, a bomb or wholesale assault on the White House. Yet in England or France, for example, the prime minister's or president's life is hardly ever in danger. There is a dearth of plots to kill them. Downing Street or the Elysée Palace are never stormed by militia. Is it because these heads of State are somehow less important? Are they less invested for us than the US president is for the Americans?

The immediate response would be to evoke the Kennedy assassination and the immense difficulty of working this through, but don't we need to look further back to colonial history here? After Charles I's execution, Cromwell took the king's place and the country was more or less under Puritan rule until the restoration of Charles II to the throne. The Puritans were the object of intense hatred and vilification, associated with both rebellion and dictatorship, and many of them, as we know, migrated to the American colonies. Their place in American history is well documented, and one might suspect that as they crossed the Atlantic they took the whole question of regicide with them. The later assassinations of Abraham Lincoln and then Kennedy might have stoked the embers of this dreadful trauma. The unthinkable crime of having executed a monarch would thus haunt American culture, with the ruler's death endlessly averted in books, TV and film.

———

An interesting variant of the standard assassination motif has emerged over the last sixty or so years. A killer is trained by a secret government agency, which, when it needs to cover its tracks, requires that the assassin himself be assassinated. In many of these sagas, the person has lost their memory or had it erased, so all they know is that someone is out to kill them. In the end, they always return home to the agency that created them and wreak revenge, as if going back to their first family.

We could read this as a narrative about repression and guilt: violence occurs, but it is not the person's responsibility. Some agency had programmed them to kill, so it is not their own fault. In other variants, the lone assassin actually starts killing for their own ends, a 'rogue' agent who must be stopped by . . . another agent. Note how the agent's hands constantly surprise him in these films, possessed of a lethal skill and dexterity he was unaware of, a feature now exploited for its comic potential in *American Ultra*. The narrative echoes curiously the old stories of a severed hand, carrying out the actions of a master yet void, apparently, of any direct responsibility.

These stories resonate with the concerns of the obsessional neurotic, who is desperate to avoid guilt and blame. One of the central mechanisms of the obsessional is undoing, reversing an action or denying that it has ever taken place. All traces of a crime have to be eradicated, a task that is never entirely possible. The person remains tortured by the violent thoughts or images that he is unable to process in this way. As in the Bourne series, the CIA do their best to negate the hero's existence, yet he keeps coming back. He can't be killed properly, and remains prey to the disturbing memories of his former acts of violence even when trying to relax on the beach.

The obsessional inhabits a strange space here. He tries to annul all traces of his violent thoughts, while at the same time perpetually awaits some transformative act of violence that will free him. That is why the characteristic action of the obsessional is non-action. He waits, and often elects as a best friend someone who, on the contrary, seems to be able to get things done, to act, as it were, in his place. The financial crash of 2008 was wonderful for some of my patients who were made redundant overnight. At last, they could definitively avoid having to make the decision to act, to quit their alienating jobs and do something meaningful like working for a charity or teaching children. This was now forced upon them, so, miraculously, they did not have to decide, to make the big choice themselves. It was made for them.

Today's regime of pseudo-activity is thus welcomed by many obsessionals. They can do their recycling, vote on talent shows, send comments on current affairs to Internet sites, and generally behave as if they are being active. But of course, these are not real activities that herald change, just elements of the new culture of pseudo-action which protect us from real dislocation. Change is awaited, but as long as the person has no part in it.

The progressive expansion of voting practices merely reinforces this. Where voting was once a political act, today we vote on just about everything, from our *X Factor* favourite to news items to service providers. On Facebook and other social media, we can click Like – and no doubt very soon Dislike – as if the ambit of human action has been reduced to a simple opposition between plus and minus. In many instances, we actually pay to do this, with the call or texting charges levied by TV shows for casting votes for contestants.

Even voting for political parties today has become a similar process of Like and Dislike, and thus robbed of any real political content. Everyone has a voice in this new pseudo-active universe, but having a voice means, of course, that no one is heard. Except in rare cases, casting a vote means absolutely nothing, apart from providing the false impression that one has made a difference. When companies contact us eagerly to find out what we really think of their services, we are again coerced into the belief that our voices count, yet, as anyone who has ever tried complaining will know, it only counts to satisfy the company's own internal administration of its staff rather than any fundamental principle of its practice. Much more honest is the advertising campaign of the fruit drink Oasis, which simply and minimally declares: 'It's summer, you're thirsty, we've got sales targets.'

Pseudo-activity works well for the obsessional, sheltering him from real change or the risks that change entails. Non-action is the rule, and discontinuities prove fascinating as long as they take place somewhere else. Many people wake every morning to listen to the news, experiencing an acute disappointment if a catastrophe or public killing has failed to occur. They would no doubt be horrified and empathize with the victims, but there is still the momentary dip when the headlines fail to disclose any sensational loss of life.

Even in Tudor and Stuart times, there was a huge market for stories of catastrophes, from storms to earthquakes to bloody crimes. These would be disseminated via broadside ballads – verse narratives printed on one side of a single sheet of paper – or news pamphlets. Seen as divine signs of spiritual disorder or providential manifestations of God's warnings to mankind, it is difficult to say, as one historian put it, if these were titillations under the pretence of religious admonition or homilies camouflaged as sensational news.

As recompense, news of the mishaps of celebrities is often appealed to, following the archaic logic of subtraction. If there is a social and economic imbalance between A and B, this can be resolved by establishing the old equation of gift economy: $A - X = B - Y$. As long as the celebrity has some problems, we can live with our own.

———

The finest architect of this space of waiting inhabited by the obsessional is Quentin Tarantino. When people refer to him, it is usually to evoke a disjointed, meandering style of speech. Conversation moves from one topic to another, with no apparent logic or regard for the gravity of the situation the characters may find themselves in. Details of hamburger preparation and philosophical questions of existence seem to have the same status. But this is not what the filmmaker really nails. On the contrary, it is not what is said but what is left out of the conversation that is his hallmark.

All his films centre on scenes where two or more people are talking and you know that a brutal act of violence is going to take place. When John Travolta and Samuel L. Jackson chat amicably to the youngsters who have stolen their boss's suitcase in *Pulp Fiction*, we know that something terrible is about to happen, just as we sense viscerally that the polite banter between Christoph Waltz and the French farmer he is questioning in the opening scene of *Inglourious Basterds* will end in slaughter. Whether it actually happens or not, Tarantino shows us how all speech does little more than keep the moment of death at bay.

This space between life and death is portrayed with unbearable accuracy in *Django Unchained*, when bounty hunter Waltz shoots plantation owner Leonardo DiCaprio. Rather than just dying, the scene prolongs the moment of DiCaprio's realization that he is dead, as he staggers slowly backwards in shock. If obsessional neurosis is at one level an attempt to defer the recognition of mortality, it means keeping these two poles as far apart as possible. In his films, Tarantino relentlessly brings them closer.

The limbo that he explores so mercilessly can also affect those who have narrowly escaped some catastrophe. The survivors of a plane crash or car accident, or those who have managed to recover from a dangerous illness, often describe the sense of unreality that ensues. If the catastrophe itself consisted of a single action – the plane hitting the ground, the car smashing into some obstacle, the virus infecting them – survival or recovery is not really a single positive action in any comparable sense. When cartoon characters get a bump on the head, the only way to get better is to receive another bump on the head. But in life, things are not like that. Recovery is usually a slow and incremental process. It is not an act. The second whack doesn't happen, leaving the person in a strange relation with mortality.

What others see as recovery can thus be just as serious as an initial accident or illness, as it deprives the person of the necessary counterpoint, the ballast, that would allow the sense of both perspective and an equivalent act of the restoration of health. If an accident *happened*, recovery didn't *happen* in the same way. We can understand how the psychoanalyst Jean Allouch could claim that mourning is perhaps more of an act than a process, and why violent acts are quite common in traditional mourning rituals. One action responds to another.

Acts of violence frequently have this curative aim, and it is not by chance that the hand has a special place here. Rather than seeing it as a mere tool or instrument for the propagation of violence, doesn't the agitation that inhabits the hand mean that it is the one part of the body that lends itself to impact and to the shock of percussion? When our hand punches and hits, as well as the effect this may produce on the one we damage, aren't we also striking the hand itself, as if to negate and expel from the body some of the internal turbulence we feel?

Many people, indeed, experience a satisfaction when they injure the hand they use to hit with. A state of unrest and unbearable tension is mediated, as the hand not only transmits pain to the one it strikes but loses pain from itself. It's why the eagerness to attack others so often precedes any perceived slight or insult: the body needs to lose something, and so we can be on the lookout for anything that will allow a focus for violence. When we can't find someone else, we can of course strike ourselves directly, a fact well known to those who self-harm.

7

Everywhere we look, people are busy with their mobiles. Social situations fail to attenuate phone use, and, for so many commentators, this spells the end of traditional human relations as we know them. What has happened, they say, to the scene of the family gathered around the table or the hearth, to people conversing with each other in public places, to the passion for the person in front of one rather than the one at the other end of the Wi-Fi?

How awful to be so distracted from the richness of human life, and yet, we could object, hasn't one of the principle aims of human life always been precisely to abstract oneself from situations of proximity with one's fellow humans, even if it's those we love? The mums and dads on their mobiles in the playgrounds aren't necessarily bad parents; they are just doing what humans unremittingly do, which is to find ways – through religion, music, craft, technology – to be somewhere else.

When car phones became a feature of popular culture in the 1980s, a recurring scene could be witnessed in TV comedy shows, films and actually on the streets: a flashily dressed man in an open-top car talks loudly into his phone, broadcasting details of his share dealings and very successful love life. We then see that the phone is not plugged in. Today an almost inverted process takes place. People text furiously in a public place, with facial expressions of great purpose, yet all they are doing is deleting old messages or typing in random letters and numbers to their phone. Their phone is not 'plugged in', as it were, yet this, for them, is the whole point. The attachment to their device here allows them to abstract themselves from the situation they are in, to be absent while being nonetheless physically present.

This process often explains the sudden receipt of unexpected texts from an old friend or acquaintance. If you are surprised and delighted to be contacted out of the blue – usually in the early evening or late at night – it may signal less an authentic outpouring of love or a reaching out to a true companion than a desperate attempt to find one more person to send a message to. The person texting you may well be in a public place, surrounded by others, with a look of fierce concentration on their face as they type out their inane greeting. Beyond the demonstration that one is important and connected to others, the texting allows an exit from the situation of proximity.

Culture has always been about finding ways to create a distance from those around us, and even today we are only allowed to address others if the aim is to

ask for directions
ask for a light
ask the time
ask for change

and never more than one of these at once. Likewise, the invitation to the most casual and welcome encounter betrays this necessity for distance. We don't say 'Let's meet' but 'Let's meet for a coffee', as if there always has to be something to mediate human relations. When James I launched his campaign against tobacco in 1604, he would ask, 'Is it not a great vanitie, that a man cannot heartily welcome his friend now, but straight they must bee in hand with *Tobacco*?' But the tobacco had to be there, just as our coffee does today, whether the word 'coffee' actually ends up referring to tea or anything else.

We noted earlier how the apparent increase in hand technology may coincide historically with the emergence of new social spaces, where the range and duration of interactions were accelerated. We could of course see the ubiquity of the fans, snuffboxes, watches and other objects as necessary markers of rank and privilege, but couldn't we also interpret them as instruments to allow social contact precisely in this sense of providing mediation, of being both there and elsewhere?

In Steele's descriptions of the coffeehouses of the eighteenth century, interaction is absolutely physical: speakers tug on listeners' buttons and neck cloths and even grab them by the collar. His plea for 'a certain interval, which ought to be preserved between the speaker and him to whom he speaks' can be understood not just in the sense of pauses in speech but as an effect of the new technology. Being in the proximity of others is possible because we have ways to abstract ourselves. As a patient put it, describing the state of unrest she was experiencing after a particular change in her life, 'The problem now is that there is nothing that takes me adequately out of myself.'

TV and Internet ads forever rehearse the scene in which a guest's arrival into a home is mediated by a hot drink, an air freshener, a cleaning product, as if the arrival of another person carries with it an intrinsic threat. The mobile technology we make so much of serves this very traditional human purpose: to abstract oneself from life. The contrast between earlier forms of relationship where people related directly to each other and contemporary ones in which they don't is pure fiction. Technologies – whether elementary or sophisticated – have always been used to generate distance.

And isn't this one of the functions of what is arguably our very first piece of technology, the bit of blanket or fabric or soft toy that Winnicott called the transitional object? Don't they allow the infant, perhaps for the first time, to be somewhere else, to abstract themselves from the body they are so entangled with?

Hands offer a first way out. Confronted with the bodily sensations that the infant has to deal with, the hand is a vector outwards. As it moves from grasping to reaching, it situates its aims outside the contours of the body. It allows a movement away from, and this is perhaps one of the reasons why hands have been so bound up with ideas of human agency. But this movement is not simple. It is not just a question of aiming outwards, since the way we use our hands shows that they form part of an arc. If they come to reach away from us, they are continually pressured to return to the body.

The hands taunt the body in so many different ways, from picking to scratching to plucking to scraping to rubbing. These activities tend to be against our conscious will, and generate shame or embarrassment when they are publicly observed. As forms of attack on our bodies, keeping our hands busy is a means of delaying or defending against them. The old warnings that the devil will seize on idle hands in fact suggest that this devil is quite personal, and will turn our hands back to our own bodies if unchecked. Work then becomes what stops the hands from returning to stimulate the body surface.

A friend who would constantly rub at a point on her nose, to such an extent that the cartilage had been damaged, explained that she welcomed her new academic job, despite the long travelling distance and arduous hours, as it would have an inhibitory effect. Teaching in front of a hall full of students meant that she would not be able to apply this morbid pressure to her nose. When we would meet socially, whenever she moved her hand towards its target, I would grab it, restraining her and thus blocking the symptomatic arc. We wondered who was more unwell: her for the compulsive action or me for compulsively blocking it.

Technology operates at one level to prevent or delay this return to the body, and can be both sophisticated, as with the smartphone, or simple, as with the many objects that people have occupied their hands with throughout history. Think of the practice of prayer or worry beads. Today, they are most often seen in the Mediterranean regions, where both men and women publicly finger beads, yet up to two thirds of the world's population use them. There are Buddhist, Muslim and Greek Orthodox bead chains, with Christianity's perhaps the last to gain popular use. There has been an increasing secularization of beads, with many users claiming they have no religious significance or link to prayer. It has been observed that, despite their long history and established use, beads are increasingly being replaced by mobiles in places where people can afford them. People sit on their porches or doorsteps texting and surfing rather than running through their rosaries.

Bead chains may have developed from early chains of knots, and are usually associated with saying prayers. The word 'bead' in fact was used first of all to refer to a spoken prayer, only later coming to designate the objects that were employed to count them. Muslim beads are used to recite the ninety-nine attributes of God, with the 100th for his name, while traditional Christian rosaries of 150 beads, divided into 'decades' of ten, are used to count prayers. The emphasis on the Paternoster prayer was gradually replaced by the Ave Maria, and this Marian stamp of the Christian rosary coincided with its most widespread use since the fifteenth century.

But why are the beads associated with counting? Lady Godiva famously bequeathed a string of threaded gems to the Benedictine priory she had founded, beads she had used 'in order that by fingering them one by one as she successively recited her prayers she might not fall short of the exact number'. The knots or beads that gained such currency were supposed to be touched like an abacus, so that the prescribed prayers could be recited without having to count aloud. But as Eithne Wilkins points out in her study of the rosary's history, bead use has never been just about counting, but about moving. Humans, as the Church understood very well, are a species that fidgets.

And this brings us to the key point. Piety is and has always been a physical practice involving the body, and the mental exercise of prayer is combined here with a physical movement of the hands and fingers. Although the involvement of the body has been more associated with Eastern religions, it is equally present in Christianity. When the early ascetics left for their solitary life of prayer, sometimes the only thing they took with them was the string of prayer beads. If we remember that until fairly recently prayers were said aloud, it shows the absolute necessity of this technology as an accompaniment to speech. Even today, as beads are progressively losing their religious function, they still operate as this combinatorial instrument.

——

There is no documented human culture in which speakers do not move their hands as they talk, and the link of hand activity to language has generated a great deal of research. Much of the work from the early 1960s on looked at how manual gesture is used to accompany speech at both a conscious and a preconscious level. Psychologists and psychiatrists studied the correlation of physical movement with pauses and punctuations in speech, hoping to find the rules of a 'bodily grammar'. Some argued that hand use will increase dramatically at the points in speech where something is difficult or impossible to symbolize. Others pursued the ambitious project of kinesics, a structural analysis of 'nonverbal' behaviour that could break gesture down into the equivalent of phonemes and sentences. More recently, there has been much research on the link of handedness to early vocalization.

The first generation of researchers agreed that hand use fell more or less into two groups. 'Object-focused' movements were closely linked to the spoken word – emphasizing, punctuating, qualifying and illustrating – while 'body-focused' movements like scratching and rubbing were not. The object-focused gestures were phased to the rhythms of speech, and fluctuations in their synchronicity could be seen as indicative of problems in verbal encoding. When movements were poorly phased with speech, the speaker may be experiencing a difficulty in representing something. But the body-focused activities seemed different: less attuned to the rhythms of speech, they would often occur after a loss – such as a separation or bereavement – as if the body was stimulating itself in response to pain and grief.

But this differentiation, shared by all of the researchers, became more complex. If the object-focused movements, like those studied by the classical rhetoricians, were linked with speaking, the body-focused ones were not totally cut off from language. If they had little link to speaking itself, what they did correlate to was precisely the experience of listening. We might use gestures knowingly and unknowingly as we try to persuade or simply communicate with our audience, but to be on the receiving end of speech also involves the body. Can anyone be spoken *to* without moving their hands?

Psychoanalysts are perhaps in a good position to assess this. After all, they spend the greater part of their day sitting mostly in silence, listening to their patients' speech. So what do they do with their hands? Curiously, the most popular image of the listening psychoanalyst ascribes a notepad to them. When the New York department store Macy's staged a window display of a psychoanalyst's office in the 1950s, complete with patient on the couch, the analyst was depicted taking notes. Yet at that time this was by no means a habitual practice, and Edmund Bergler would swiftly publish an article about the myth of the note-taking analyst. Freud had advised against it, and in fact, a survey of analytic literature up to the present day shows that the single most common recorded practice for the listening psychoanalyst is not note-taking but knitting.

Freud's daughter Anna was famous for this, and she would even suggest that her students be taught weaving as part of their psychoanalytic training, in order to 'balance' the work of listening. Where Freud's study at the house in London where he spent the last year of his life is filled with antiquities, Anna's room contains a huge loom, testimony to the place that manual activity had for her. Her hands, she said, had served her well, absorbing what her father called her 'passionate excesses', and when they refused to knit after a debilitating stroke in 1982, she commented, 'Look at what that hand did, it is angry because I controlled it for so long.'

Freud himself kept his hands busy with the materials of smoking, and he would also often lick his gem ring when listening, a habit that was apparently once common. Likewise, the many antiquities that adorn his consulting room were not simply objects of his or his patients' gaze. As he sat behind the couch, his hands would turn over and fondle these small and sometimes fragile figurines and amulets.

Doodling is also very frequent among analysts, and is no doubt closely linked to the experience of listening. We doodle, after all, perhaps most often when we are listening to someone on the phone or in a meeting. When a London magazine invited entries for a doodling competition in the 1920s, psychiatrists used the 9,000 submissions for their own study. What they found was that most people doodled not when doing nothing or at a loose end but, on the contrary, when they were listening to the radio. Once again, it shows that we should question the popular image of a family huddled around the wireless in the good old days of human relationships. Even then, people had to keep their hands busy. And here we see too the association of hand activity with listening, as if the experience of speech needs to be somehow embodied.

To make a series of works, the Brazilian artist Rivane Neuenschwander visited bars, places where humans talked and listened to each other, collecting the small, inconspicuous remainders of the exchange of speech: fragments of cardboard folded or rolled up, matchsticks split and twisted, coasters grooved and bent. She called these objects *Speech Acts*, to highlight the fact that they were the by-products of conversation. As her work showed so clearly, language does not exist in any disembodied state but demands incarnation, with the hands constantly manipulating and shaping.

A patient explained how when she experienced a high or a low in her emotional life, she had to provoke pain in her body to mark it. This was not, she said, to annul it or take her mind away from it, but simply to generate the corresponding bodily state, a 'kind of balance'. 'People,' she said, 'need to balance psychical pain with physical pain.' This, for her, was absolutely normal, and we might wonder whether it is, in fact, a principle that guides the logic of our hands.

As we are worked by language, as our thoughts pull and push us in different, often painful directions, do we use our hands to try to generate this balance? And if words require an embodiment via the use of our hands, is its ultimate form the act of writing?

—

Hand movements can be orchestrated and contrived, but perhaps they are ultimately less an accessory or instrument of speech than a part of speech itself. They form part of speaking – and, crucially, of listening – as words take hold not only of our minds but of our bodies as well. We could note here how blind

99

people use hand gestures with the same frequency and the same form as the sighted, as if gesture emerged, as the psychologist Susan Goldin-Meadow put it, not only for the sake of others but also for ourselves.

As we speak and are spoken to, we do our best to embody words, whether through gestural movement, or fidgeting, or doodling, or writing. We could remember here that if Christianity has always emphasized Logos, it has also relentlessly transmitted the idea of incarnation, of how the word demands embodiment. This is not an abstract question, but one which concerns everyone, believer or not.

Writing, indeed, can be considered a form of gesture, as the body is involved in making a mark or inscription. As we see the passengers on the tube or bus frantically texting, scrolling and clicking, shouldn't we understand this as a kind of generalized practice of writing, made all the more urgent by the invasive presence of words, as they are beamed at us day and night in our new digital world?

In his groundbreaking study of gesture, David Efron compared the body language of Southern Italians and Eastern Jews just off the boat in New York with those who had lived there already for several years. Writing in the 1930s, he aimed to challenge the prevailing racist theories that associated religious and national characteristics with behavioural traits. Gathering his material in streets, parks, markets, synagogues, social gatherings and restaurants, he found that gesture did indeed change in the new environment, and, remarkably, that the planes of motion involved in gesticulation suggested a kind of script.

Both the Southern Italians and the Eastern Jews who had recently arrived used their hands quite differently from their assimilated counterparts, with the Italian gestures tending to follow a single direction till the hand movement was completed, within a spherical surface plane. In contrast, the Jewish gesticulation switched from one plane of motion to another, using an angular change in direction, to give zigzag lines, which, if reproduced on paper, gave the image of 'an intricate embroidery'. They were described by one observer as 'like entangled thread', and of having 'a knitting-like quality, very difficult to describe'. The diagrams in Efron's book in fact show the form of a trefoil knot, something we would associate with mathematics rather than human speech.

Writing and the body seem to converge here, and we could note how there has always been an association between the idea of weaving and knitting and that of speaking and storytelling. Hans Christian Andersen supposedly learnt many of his stories by listening to them in a weaving room, and textiles have historically often been made by groups who chattered as they worked. We speak of a fate or destiny being woven or spun, and the goddesses responsible are depicted as spinners and weavers. Just as a yarn is a story, so the action of the hands accompanies the action of speech.

In his later years, Jacques Lacan introduced knot theory into psychoanalysis, arguing that it provided the most rigorous model available to map the human psyche. He would carry string around with him in his pockets, knotting and unknotting at all places and times, from his consulting room to his public seminar to the cafés and restaurants he frequented. We might guess here that beyond the theoretical advantage of knot theory, Lacan was interested in the actual practice of knotting, and it was this that kept his hands busy for so many hours.

It is curious to observe that the start of Freud's analytic career is often characterized by a renunciation of the use of the hands, yet now, towards the end of his own, Lacan brings them back. In his early work with hypnosis, Freud would hold his hand to the patient's forehead, as he elicited thoughts and memories. With the psychoanalytic technique of free association, the hands were definitively abandoned. Yet Lacan's activity of knotting shows the convergence of analytic and manual work: the knots were hardly an abstraction for him, but a constant daily activity, and it is this real, embodied dimension that we should reflect on.

—

This push to embodiment, to incarnation, and perhaps to writing, illuminates why hand activity so often takes place in conjunction with something else. There is listening AND doodling, knitting AND talking, praying AND manipulating beads. The real question here is the AND. It is a pity that despite an early interest from Freud and his student Sándor Ferenczi in the fusions and intermixing of bodily drives, psychoanalysis never really took up this problem. While commentators endlessly point out that the key to Freud's patient Dora's phantasy life was the image of her sucking her thumb and simultaneously tugging her brother's earlobe, they have plenty to say about the sucking and the tugging but nothing about the AND in between them.

Yet even the most cursory consideration of the main examples of bodily drives shows that there is always an AND. The voyeur may be frozen in the act of looking, but he is also always doing something else with his hands. The infant feeding at the breast is also simultaneously using her hands and fingers to clasp, grip, fan and touch. And as we grow older, we might bite or pick our nails or scratch or rub as we read or watch TV. Many people, indeed, are only able to work if they are simultaneously pulling or scratching or curling. Why isn't one activity enough? Why can't sucking or reading or watching suffice for us?

There is now a massive industry catering for the human practice of snacking and grazing. TV on its own is not the answer, as the hand has to move food continually to replenish the mouth. Although the biting and chewing and the sugar and salt have their importance, we should see this as another example of keeping hands busy and, ultimately, of the 'morbid alliance' of hand and mouth. Erasmus noted that 'Continuous eating should be interrupted now and again with stories. Some people eat or drink without stopping not because they are hungry or thirsty but because they cannot otherwise moderate their gestures, unless they scratch their head, or pick their teeth, or gesticulate with their hands, or play with their dinner knife, or cough, or clear their throat, or spit.'

We could wonder whether the watching is the basic activity here or the eating. Indeed, the difficulties that Erasmus pointed out are just as present today as they were in the sixteenth century. A man described his anxiety on learning that the dinner service at the country hotel he was staying at would involve a buffet-style meal. Rather than having courses served, he would have to select and, crucially, time his starter, main course and dessert himself, yet he had no idea what to do in between courses. How long should he wait before getting up? What should he do in the intervals between courses?

Goffman had observed many years ago how difficult it was for someone to eat alone, since direct involvement with food is not generally allowed by society. You have to have a higher purpose, like a newspaper or a book, or, today, urgent and important text messages that require deliberation and response. The same repression of an interest in food applies to pornography. In the days when girlie magazines adorned the top shelves of newsagents, a sociologist found that men would flick through them without ever stopping at a particular page, as if attention had to be evenly distributed. Facial expressions would show furrowed concentration, as if they were searching for some important and transcendent information that required systematic page turning.

The interruptions advocated by Erasmus separate the eater from their food, just as the course divisions of a conventional restaurant establish punctuation and interval. Much of everyday life has this same aim, to create a distance from the appetites that overwhelm us, shared by mouth and hand. Reality shows such as *Gogglebox*, which documents a range of British families watching TV, testify to this unrelenting bifocality: however gripped they may be by what is onscreen, they are always also doing something with their hands. Discussions of screen time tend to miss this key fact, that watching means that hands have to be kept occupied, yet we need look no further than the remote control.

People in the US and the UK watch an average of four hours of TV every day. During this time, they may frequently change channels, and this has been linked with a rise in attention deficit problems, failure to focus, and a general sense of distraction. But we could see channel hopping not simply as a sign of attention difficulties but of hand use. All families fight over the remote control, and if it is at one level a symbol of domestic power, it is also an instrument to hold, to fiddle with, to press, to click.

Screen time is always also hand time, and just as the remote became a hand technology so quickly, so today as people watch TV they are simultaneously on their mobiles and tablets. This is no doubt the reason why the puzzlingly vapid yet ever-present aeroplane magazine will never be entirely replaceable by a screen, or at least not until all mobiles are usable during flight time.

—

What happens when the hands are kept from their business? If a child is separated from their Lego or loom bands, or an adult from their mobile or console, very soon the body takes its place. Hands scratch, rub, pluck, press and pull. The overflowing agitation of the hands finds its way back to the body, a return that hand technology defers and inhibits. There is a circuit here that pulsates with a ferocious energy. It is also no doubt one of the reasons for the ultimate failure of nicotine patches: although they supply the chemical, they can never replace the activity of holding and manipulating the cigarette.

Manual stimulation of the body is often described as comforting, but it can also involve a modulation of pain. Describing her extraction of scalp hairs, a woman explained that 'I want to get to the same feeling again, that exact feeling when the hair comes out, like a needle.' However many hairs she removed, it was only rarely that she could refind the sensation she sought, the match that Freud saw as central to autoerotic activities. If the plucking aimed at a pain, it was a pain that matched another, that created an identity of perception.

This refinding is about as close as we can get to personal identity. When philosophers discuss this topic, they tend to assume that people go around with a stable sense of who they are, and then try to investigate what they take to be some sort of underlying substrate. But in my experience, hardly anyone has a sense of who they are, but they may feel, at certain punctual moments, that they exist, often through the production of pain. Psychoanalytic theories of how we come to inhabit our bodies tend to neglect this feature: that pain sharpens and defines bodily boundaries, as we see from a variety of practices.

Take the most anodyne example of a man entering a café or restaurant to meet friends or a date. The same thing will happen with an almost comical regularity. As he crosses the room, he will touch his face or head. It won't be a scratch or a rub, just a touch to the chin or the nose or the neck. Actors perform an identical action when they are trying to act like real people: they touch their faces. Even a genius like Tom Hardy does it all the time.

These curious self-anointments tend to take place when others are present and the person feels watched or, at the least, seen. As the gaze of the other is imagined to turn to them, they touch their own body, as if to affirm that they are there, and to disarm the potentially negative gaze. It is perhaps no accident that in many cultures, a representation of the hand is used to ward off the evil eye. As Goffman pointed out many years ago, the social self here is not a given but an act to be performed in front of an audience. If we feel the presence of this audience, the discreet moment of self-stimulation when crossing the room anchors the person in their body.

This fleeting bodily sensation is at one end of a spectrum, with forms of self-harm at the other. Both aim to make the person exist through a return to the body, with the physical feeling we generate in ourselves acting as our point of consistency, what we go back to. It is difficult here to separate self-harm from the self. As one man put it, 'It's only when I am cutting myself that I can feel me, separate from others, the only way I can reach myself.' For Terri Cheney, whom we quoted earlier, if her body was experienced as her mother's creation, cutting it still allowed a route back to herself. In the words of another patient, 'It's the only time I can feel that I exist.'

The cut here is an autonomous act, and the cutter the sole agent. As well as permitting an idea of agency, it can introduce a rhythm, a sense of before and after, crucial if the person feels caught in a relentless and never-ending experience of numbness or anxiety. Could this be linked to how we spend so much of our time at the dawn of life touching our own bodies with our hands, and, indeed, why the infant is so fascinated with the moment when one hand can touch the other? As we grow older, is pain the only way that we can reconnect with ourselves?

⎯

This brings us back in a way to our point of departure. Rather than focusing on our alienation in the digital world, I wanted to emphasize how keeping our hands busy has always been a central human activity. Life has never just been about connecting, but also about disconnecting. Phones, computers and tablets allow an abstraction from our proximity to others and the demands that this involves. As mediators, they allow us to be somewhere else. But ultimately, if they form part of a kind of manual circuit, which moves back towards the stimulation of the body, they do indeed alienate us. This alienation may keep us from ourselves, and pain can bring us back, when the hands return to the body. But it is alienation in a positive and not simply a negative sense, fulfilling a necessary function in human life, helping us to localize and operate with the tension and agitation that inhabit us.

It is for this reason that the many projects designed to distance our hands from the new technology are likely to fail. We have already seen the poor response to Google Glass, and the prospect of communicating with our computers solely through the voice may seem exciting and attractive, but it will deprive us of perhaps the most important role that this technology has: to allow our hands to tap, type, click and scroll. Language, as we have seen, needs to be embodied, and the omnipresence of the Word requires that our hands stay busy turning this word into flesh.

Apple, once again, have understood something here that their competitors have missed. If you phone them, you are greeted by an automated system that handles calls. It asks you what you want, and when you tell them, there is a little pause before it tells you what to do next. And in that little pause, you hear a sound, the sound of manual typing, as if your message were being typed by human hand. Of course, it is a computer-generated sound, and it bears no relation at all to what you have just said, but it shows that in that gap, in that space between question and response, between speech and its reply, we need to know that hands are still there.

Notes

p. 2   Apple, see William Merrin, 'The rise of the gadget and hyperludic media', *Cultural Politics*, 10 (2014), pp. 1–20. Changes in hands, see Frank R. Wilson, *The Hand: How Its Use Shapes the Brain, Language, and Human Culture* (New York: Pantheon, 1998); and John Napier, *Hands* (New York: Pantheon, 1983). Overbite, see Bee Wilson, *Consider the Fork: A History of How We Cook and Eat* (London: Penguin, 2012), pp. 100–106.

p. 3   Hands as gesture, see Jean-Claude Schmitt, *La Raison des gestes dans l'Occident médiéval* (Paris: Gallimard, 1990); Jan Bremmer and Herman Roodenburg (eds), *A Cultural History of Gesture* (Oxford: Oxford University Press, 1991); and Michael J. Braddick (ed.), *The Politics of Gesture: Historical Perspectives* (Oxford: Oxford University Press, 2009).

pp. 5–6   On the image of the hand, see Marjorie O'Rourke Boyle, *Senses of Touch: Human Dignity and Deformity from Michelangelo to Calvin* (Leiden: Brill, 1998); Elizabeth D. Harvey (ed.), *Sensible Flesh: On Touch in Early Modern Culture* (Philadelphia: University of Pennsylvania Press, 2003); Claire Richter Sherman (ed.), *Writing on Hands: Memory and Knowledge in Early Modern Europe* (Seattle: University of Washington Press, 2001); Katherine Rowe, 'God's Handy Worke', in David Hillman and Carla Mazzio (eds), *The Body in Parts: Fantasies of Corporeality in Early Modern Europe* (New York: Routledge, 1997), pp. 285–309; and Michael Neill, ' "Amphitheaters in the body": Playing with hands on the Shakespearian stage', *Shakespeare Survey*, 48 (1996), pp. 23–50. Divine hand, see Bernhard Scholz, ' "Ownerless Legs or Arms Stretching from the Sky": Notes on an Emblematic Motif ', in Peter M. Daly (ed.), *Andrea Alciato and the Emblem Tradition: Essays in Honor of Virginia Woods Callahan* (New York: AMS Press, 1989), pp. 249–83; and Paul

McPharlin, *Roman Numerals, Typographic Leaves and Pointing Hands: Some Notes on Their Origin, History, and Contemporary Use* (New York: The Typophiles, 1942). Aristotle's praise, see *Parts of Animals* 687a 10. Hand in oratory, see Gregory Aldrete, *Gestures and Acclamations in Ancient Rome* (Baltimore: Johns Hopkins, 1999); and Leanne Bablitz, *Actors and Audience in the Roman Courtroom* (London: Routledge, 2007), pp. 187–92.

pp. 7–8 See Sergio Della Sala et al, 'The anarchic hand: A fronto-medial sign', *Handbook of Neuropsychology*, 9 (1994), pp. 233–55. Throttling, etc., see Gordon Banks et al, 'The alien hand syndrome', *Archives of Neurology*, 46 (1989), pp. 456–9; Gary Goldberg and Karen K. Bloom, 'The alien hand sign', *American Journal of Physical Medicine and Rehabilitation*, 69 (1990), pp. 228–38; and Alan J. Parkin, 'The Alien Hand', in Peter W. Halligan and John C. Marshall (eds), *Method in Madness: Case Studies in Cognitive Neuropsychiatry* (Hove: Psychology Press, 1996), pp. 173–83. Kurt Goldstein, 'Zur Lehre von der motorischen Apraxie', *Journal für Psychologie und Neurologie*, 11 (1908), pp. 169–87 and 270–83. Like a child, see Rachelle Smith Doody and Joseph Jankovic, 'The alien hand and related signs', *Journal of Neurology, Neurosurgery and Psychiatry*, 55 (1992), pp. 806–10. Differentiating, see Clelia Marchetti and Sergio Della Sala, 'Disentangling the alien and anarchic hand', *Cognitive Neuropsychiatry*, 3 (1998), pp. 191–207; and Leonardo Caixeta et al, 'Alien hand syndrome in AIDS', *Dementia and Neuropsychologia*, 1 (2007), pp. 418–21. Extra hand, see Riitta Hari et al, 'Three hands: Fragmentation of human bodily awareness', *Neuroscience Letters*, 240 (1998), pp. 131–4; and F. Aboitiz et al, 'The alien hand syndrome: Classification of forms reported and discussion of a new condition', *Neurological Sciences*, 24 (2003), pp. 252–7. For the related motif of a hand guiding a hand in art, see Moshe Barasch, *Giotto and the Language of Gesture* (Cambridge: Cambridge University Press, 1987); and Richard E. Spear, *The 'Divine' Guido: Religion, Sex, Money and Art in the World of Guido Reni* (New York: Yale University Press, 1997).

p. 10 Camp survivors, Nathalie Zajde, *Guérir de la Shoah* (Paris: Odile Jacob, 2005) and *Enfants de survivants* (Paris: Odile Jacob, 1993).

p. 12 Aristotle, touch needed for life, *On the Soul*, 435ab. See Elizabeth D. Harvey, 'The portal of touch', *American Historical Review*, 116 (2011), pp. 385–400.

p. 13 Shared ideology, see Nicole Dawkins, 'Do-it-yourself: The precarious work and postfeminist politics of handmaking in Detroit', *Utopian Studies*, 22 (2011), pp. 261–84. Violence of the market, see Slavoj Žižek, *Violence* (London: Profile, 2008).

p. 15 20%, see A. Korner and H. Kraemer, 'Individual Differences in Spontaneous Oral Behaviour in Neonates', in J. F. Bosma (ed.), *Third Symposium on Oral Sensation and Perception* (Springfield: Charles Thomas, 1972), pp. 335–46; Philippe Rochat, 'Hand–Mouth Coordination in the Newborn', in G. J. P. Savelsbergh (ed.), *The Development of Coordination in Infancy* (Amsterdam: Elsevier Academic Press, 1993); Philippe Rochat and Stefan Senders, 'Active Touch in Infancy: Action Systems in Development', in M. J. Weiss and P. R. Zelazo (eds), *Infant Attention: Biological Constraints and the Influence of Experience* (New Jersey: Ablex, 1991), pp. 412–42; and Burton White et al, 'Observations on the development of visually-directed reaching', *Child Development*, 35 (1964), pp. 349–64.

pp. 16–17 Jean Piaget, *The Origins of Intelligence in Children* (New York: International Universities Press, 1952). René Spitz, 'Digital extension reflex', *Archives of Neurology and Psychiatry*, 113 (1950), pp. 467–70. No rooting if their own hands, George Butterworth and Brian Hopkins, 'Hand–mouth coordination in the new-born baby', *British Journal of Developmental Psychology*, 6 (1988), pp. 303–14. Hand and mouth, see H. M. Halverson, 'Studies of the grasping responses of early infancy', III, *Journal of Genetic Psychology*, 51 (1937), pp. 425–48, and 'Infant sucking and tensional behaviour', *Journal of Genetic Psychology*, 53 (1938), pp. 365–430. René Spitz, 'The primal cavity: A contribution to the genesis of perception', *Psychoanalytic Study of the Child*, 10 (1955), pp. 215–40. Selma Fraiberg, *Insights from the Blind* (New York: Basic Books, 1977); and Edna Adelson and Selma Fraiberg, 'Mouth and Hand in the Early Development of Blind Children', in J. F. Bosma (ed.), *Third Symposium on Oral Sensation and Perception*', op. cit., pp. 420–30. Arnold Gesell, *Infant Development* (London: Hamish Hamilton, 1952).

p. 18 Colwyn Trevarthen, 'Biodynamic Structures, Cognitive Correlates of Motive Sets and the Development of Motive in Children', in W. Prinz and A. F. Sanders (eds), *Cognition and Motor Processes* (Berlin: Springer, 1984), pp. 327–51. W. Kessen, 'Sucking and Looking: Two Organized Congenital Patterns of Behaviour in the Human Newborn', in

H. W. Stevenson et al, *Early Behaviour* (New York: Wiley, 1967); and Albrecht Peiper, *Die Eigenart der kindlichen Hirntätigkeit* (Leipzig: Thieme, 1949). Twenty weeks, see Marc Jeannerod, *The Neural and Behavioural Organization of Goal-Directed Movements* (Oxford: Clarendon, 1988). Andrew N. Meltzoff, 'Molyneux's Babies: Cross-Modal Perception, Imitation and the Mind of the Preverbal Infant', in Naomi Eilan et al, *Spatial Representation: Problems in Philosophy and Psychology*' (Oxford: Blackwell, 1993), pp. 213–34; and Andrew N. Meltzoff and Richard W. Borton, 'Intermodal matching by human neonates', *Nature*, 282 (1979), pp. 403–4.

p. 19 Jerome S. Bruner, 'Eye, Hand and Mind', in David Elkind and John Flavell (eds), *Studies in Cognitive Development: Essays in Honor of Jean Piaget* (Oxford: Oxford University Press, 1969), pp. 223–35.

p. 20 Selma Fraiberg, *Insights from the Blind*, op. cit.

pp. 22–3 'Bubby Pot', see Arnold Gesell and Frances L. Ilg, *Child Development* (New York: Harper, 1943); and Valerie Fildes, *Breasts, Bottles and Babies: A History of Infant Feeding* (Edinburgh: Edinburgh University Press, 1986). Freud, *Three Essays on the Theory of Sexuality* (1905), *Standard Edition*, vol. 7, pp. 130–253. Donald W. Winnicott, 'Transitional Objects and Transitional Phenomena' (1953), in *Playing and Reality* (Harmondsworth: Penguin, 1971), pp. 1–30. Competition, see Willi Hoffer, 'Mouth, hand and ego-integration', *Psychoanalytic Study of the Child*, 4 (1949), pp. 49–56, and also his 'Oral aggressiveness and ego development', *International Journal of Psychoanalysis*, 31 (1950), pp. 156–60, and 'Development of the body ego', *Psychoanalytic Study of the Child*, 5 (1950), pp. 18–23.

p. 24 Freud, *The Interpretation of Dreams* (1899), *Standard Edition*, vol. 5, pp. 566–7.

pp. 25–6 Terri Cheney, *The Dark Side of Innocence: Growing up Bipolar* (New York: Atria, 2011), p. 142. Sites of exchange, see Karin Stephen, *The Wish to Fall Ill: A Study of Psychoanalysis and Medicine* (Cambridge: Cambridge University Press, 1960); and Darian Leader and David Corfield, *Why Do People Get Ill?* (London: Hamish Hamilton, 2007), pp. 164–205. René Spitz, 'The primal cavity', op. cit.

p. 28 Dangling, see Albrecht Peiper, *Die Hirntätigkeit des Säuglings* (Berlin: Springer, 1928). Imre Hermann, 'Clinging–going-in-search: A contrasting pair of instincts and their relation to sadism and masochism' (1935),

The Psychoanalytic Quarterly, 45 (1976), pp. 5–36. See also Livia Nemes, 'Importance de la théorie de l'agrippement dans la psychologie du développement et dans la psychothérapie d'enfant', *Le Coq-héron*, 188 (2007), pp. 73–9.

p. 29 Michael Balint, 'Early developmental states of the ego', *International Journal of Psychoanalysis*, 30 (1939), pp. 265–73.

p. 34 Aristotle, *Parts of Animals*, 687b, 690a. Freud, *Beyond the Pleasure Principle* (1920), *Standard Edition*, vol. 18, pp. 7–64.

p. 39 Extraction, see Darian Leader and David Corfield, *Why Do People Get Ill?*, op. cit.

p. 41 On drive and desire, see Slavoj Žižek, *The Ticklish Subject: The Absent Centre of Political Ontology* (London: Verso, 2000), pp. 290–306.

pp. 42–3 Keeping busy, see Monika Fludernik and Miriam Nandi (eds), *Idleness, Indolence and Leisure in English Literature* (London: Palgrave, 2014); and Robert W. Malcomson, *Popular Recreations in English Society, 1700–1850* (Cambridge: Cambridge University Press, 1973). Norbert Elias, *The Civilizing Process, Vol. I: The History of Manners* (1939) (Oxford: Blackwell, 1969) and *The Civilizing Process, Vol. II: State Formation and Civilization* (1939) (Oxford: Blackwell, 1982); and Barbara Rosenwein, 'Worrying about emotions in history', *American Historical Review*, 107 (2002), pp. 821–45. On 'civilizing', see also Joan Wildeblood and Peter Brinson, *The Polite World: A Guide to English Manners and Deportment from the Thirteenth to the Nineteenth Century* (Oxford: Oxford University Press, 1965); C. Dallett Hemphill, *Bowing to Necessities: A History of Manners in America, 1620–1860* (Oxford: Oxford University Press, 1999); Paul Lacroix, *Manners, Customs and Dress During the Middle Ages and the Renaissance Period* (1876) (New York: Skyhorse, 2013); Georges Vigarello, 'The Upward Training of the Body from the Age of Chivalry to Courtly Civility', in Jonathan Crary et al (eds), *Fragments for a History of the Human Body*, vol. 2 (New York: Zone Books, 1989), pp. 149–99.

pp. 44–5 Gail Kern Paster, 'Nervous Tension: Networks of Blood and Spirit in the Early Modern Body', in David Hillman and Carla Mazzio (eds), *The Body in Parts*, op. cit., pp. 107–25. See also her *The Body Embarrassed: Drama and the Disciplines of Shame in Early Modern England* (Ithaca: Cornell University Press, 1993). On humoral theory and its legacies, see Nancy Siraisi, *Medieval and Early Renaissance Medicine* (Chicago:

University of Chicago Press, 1990); and Michael Schoenfeldt, *Bodies and Selves in Early Modern England: Physiology and Inwardness in Spenser, Shakespeare, Herbert and Milton* (Cambridge: Cambridge University Press, 1999).

p. 46 Gordon Allport, *Becoming* (New York: Yale University Press, 1955), p. 43.

pp. 47–8 Evaluating gesture, see Jan Bremmer and Herman Roodenburg (eds), *A Cultural History of Gesture*, op. cit.; and Michael J. Braddick (ed.), *The Politics of Gesture*, op. cit. Jean-Claude Schmitt, 'Between text and image: The prayer gestures of Saint Dominic', *History and Anthropology*, 1 (1984), pp. 127–62; Etienne Robo, 'Pray with your hands', *Worship*, 33 (1958), pp. 14–18; John Craig, 'Bodies at Prayer in Early Modern England', in Alec Ryrie and Natalie Mears (eds), *Worship of the Parish Church in Early Modern England* (Aldershot: Ashgate, 2013), pp. 173–96; and Gerhart Ladner, 'The Gestures of Prayer in Papal Iconography of the Thirteenth and Early Fourteenth Centuries', in Sesto Prete (ed.), *'Didascaliae': Studies in Honour of Anselm M. Albareda* (Rome, 1961), pp. 246–75. Re eyes, see J. A. Burrow, *Gesture and Looks in Medieval Narrative* (Cambridge: Cambridge University Press, 2002).

p. 49 Objects, see Arjun Appadurai, *The Social Life of Things: Commodities in Cultural Perspective* (Cambridge: Cambridge University Press, 1986); Tara Hamling and Catherine Richardson, *Everyday Objects: Medieval and Early Modern Material Culture and Its Meanings* (Farnham: Ashgate, 2010); Cordelia Beattie et al (eds), *The Medieval Household in Christian Europe, 850–1550* (Turnhout: Brepols, 2003); Maxine Berg and Helen Clifford (eds), *Consumers and Luxury: Consumer Culture in Europe 1650–1850* (Manchester: Manchester University Press, 1999); and Alan Dessen and Leslie Thomson (eds), *A Dictionary of Stage Directions in English Drama, 1580–1642* (Cambridge: Cambridge University Press, 1999). On new technology and its distribution, see Patricia Fumerton and Simon Hunt (eds), *Renaissance Culture and the Everyday* (Philadelphia: University of Pennsylvania Press, 1998); Neil McKendrick et al, *The Birth of a Consumer Society: Commercialization of Eighteenth-Century England* (London: Europa, 1982); and Joan Thirsk, *The Development of a Consumer Society in Early Modern England* (Oxford: Clarendon, 1978). On gloves and fans, see Pearl Binder, *Muffs and Morals* (London: Harrap, 1953); Marjorie O'Rourke Boyle, 'Coquette at the Cross? Magdalen in the Master of the Bartholomew Altar's deposition at the Louvre', *Zeitschrift für Kunstgeschichte*, 59 (1996), pp. 573–7; Kate Smith, 'In her hands: Materialising distinction in Georgian Britain', *Cultural and Social History*, 11 (2014), pp. 489–506. On hankies, see Margarete Braun-Ronsdorf, *The History of the Handkerchief* (Leigh-on-Sea: F. Lewis, 1967); and Ian Smith, 'Othello's Black Handkerchief', in Lena Cowen Orlin (ed.), *Othello: The State of Play* (London: Bloomsbury, 2014), pp. 95–120.

pp. 50–52 Evelyn Welch, 'Art on the edge: Hair and hands in Renaissance Italy', *Renaissance Studies*, 23 (2008), pp. 240–68; Will Fisher, 'The Renaissance beard: Masculinity in early modern England', *Renaissance Quarterly*, 54 (2001), pp. 155–87; and the essays on hair in *Eighteenth-Century Studies*, 38 (2004). Snuff and tobacco, see J. T. McCullen, 'Tobacco: A recurrent theme in eighteenth-century literature', *Bulletin of the Rocky Mountain Modern Language Association*, 22 (1968), pp. 30–39; Roy Porter and Mikuláš Teich (eds), *Drugs and Narcotics in History* (Cambridge: Cambridge University Press, 1995); and Marcy Norton, *Sacred Gifts, Profane Pleasures: A History of Tobacco and Chocolate in the Atlantic World* (Ithaca: Cornell University Press, 2008). Re the English, this apparent absence is in fact historically variable. If Adam Smith commented on their lack of gesticulation, Dr Johnson could be known to actually seize people's hands and hold them down to dampen this propensity to movement. And where the psychologist Sidney Jourard, watching pairs of people conversing in coffee shops in the 1960s, could find that the French touched each other 110 times in one hour and the English zero times, in the late fifteenth and sixteenth centuries the English were known for their enthusiastic greeting kisses and, later, for their handshaking. In Rabelais's debate between Panurge and 'a great English scholar', the dialogue is indeed conducted not with words but entirely through gesture. See J. A. Burrow, *Gestures and Looks in Medieval Narrative*, op. cit; Jan Bremmer and Herman Roodenburg (eds), *A Cultural History of Gesture*, op. cit; and Sidney Jourard, 'An exploratory study of body-accessibility', *British Journal of Social and Clinical Psychology*, 5 (1966), pp. 221–3. Private life, see Paul Veyne and Georges Duby (eds), *A History of Private Life*, 5 vols (Cambridge, Mass.: Harvard University Press, 1987–98). Tobacco and punctuation, see William Davies, 'What have we lost in the shift from cigarettes to smartphones?', *Open Democracy*, 12 March 2015.

p. 53 J. C. Flügel, *The Psychology of Clothes* (London: Hogarth, 1930), p. 186, fn. 2. Ariane Fennetaux, 'Women's pockets and the construction of privacy in the long eighteenth century', *Eighteenth-Century Fiction*, 20 (2008), pp. 307–34; Barbara Burman, 'Pocketing the difference: Gender and pockets in nineteenth-century Britain', *Gender and History*, 14 (2002), pp. 447–69; Christopher Todd Matthews, 'Form and deformity: The trouble with Victorian pockets', *Victorian Studies*, 52 (2010), pp. 561–90; and Barbara Burman and Seth Denbo, *Pockets of History: The Secret Life of an Everyday Object* (Southampton: University of Southampton, 2007). Erving Goffman, *Behavior in Public Places: Notes on the Social Organization of Gatherings* (New York: Free Press, 1963), p. 133.

p. 54 Adam Smith, *The Theory of Moral Sentiments* (1759) (Oxford: Oxford University Press, 1976), pp. 180–83. Ariane Fennetaux, 'Toying with Novelty: Toys, Consumption and Novelty in Eighteenth-Century Britain', in Bruno Blondé et al, *Fashioning Old and New* (Turnhout: Brepols, 2009), pp. 17–28.

pp. 55–7 Idleness and work, see Hans-Joachim Voth, *Time and Work in England, 1750–1830* (Oxford: Clarendon, 2000); Hugh Cunningham, *Leisure in the Industrial Revolution* (London: Taylor and Francis, 1980); Monika Fludernik and Miriam Nandi (eds), *Idleness, Indolence and Leisure in English Literature*, op. cit.; and Sarah Jordan, *The Anxieties of Idleness: Idleness in Eighteenth-Century British Literature and Culture* (Lewisburg: Bucknell University Press, 2010). Soldiers, see Ruth Kenny, Jeff McMillan and Martin Myrone, *British Folk Art* (London: Tate Publishing, 2014); and Emmanuel Cooper, *People's Art: Working-Class Art from 1750 to the Present Day* (Edinburgh: Mainstream, 1994). On knitting and embroidery, see Anne Macdonald, *No Idle Hands: The Social History of American Knitting* (New York: Ballantine, 1988); Joanne Turney, *The Culture of Knitting* (Oxford: Berg, 2009); Rozika Parker, *The Subversive Stitch: Embroidery and the Making of the Feminine* (London: The Women's Press, 1984); and J. C. Turner and P. van de Griend, *History and Science of Knots* (Singapore: World Scientific, 1996). Dorcas, see Anne Macdonald, *No Idle Hands*, op. cit., p. 341. Boys, ibid., p. 235.

p. 58 Quotes from ibid., pp. 341–2 and 360.

pp. 59–60 Thomas W. Laqueur, *Solitary Sex: A Cultural History of Masturbation* (New York: Zone Books, 2003). Aleksandr Solzhenitsyn, *The First Circle*, quoted in Elaine Scarry, *The Body in Pain: The Making and Unmaking of the World* (Oxford: Oxford University Press, 1985), p. 48. See also the discussion in Irwin M. Marcus and John J. Francis (eds), *Masturbation: From Infancy to Senescence* (New York: International Universities Press, 1975); and Milton R. Sapirstein, *Paradoxes of Everyday Life: A Psychoanalyst's Interpretations* (New York: Random House, 1955), pp. 146–72.

p. 61 Yale, see H. M. Halverson, 'Infant sucking and tensional behaviour', *Journal of Genetic Psychology*, 53 (1938), pp. 365–430. Karin Stephen, *The Wish to Fall Ill*, op. cit., p. 186.

p. 65 René Spitz, 'Autoerotism re-examined: The role of early sexual behavior patterns in personality formation', *Psychoanalytic Study of the Child*, 17 (1962), pp. 283–315; and Herman Roiphe and Eleanor Galenson, *Infantile Origins of Sexual Identity* (New York: International Universities Press, 1981).

p. 66 On geography of the body, see Selma Fraiberg, 'Tales of the discovery of secret treasure', *Psychoanalytic Study of the Child*, 9 (1954), pp. 218–41.

p. 68 Margaret Schlauch, *Chaucer's Constance and Accused Queens* (New York: New York University Press, 1927); and Marina Warner, *From the Beast to the Blonde: Fairy Tales and Their Tellers* (London: Chatto and Windus, 1994).

pp. 71–2 Herman Roiphe and Eleanor Galenson, *Infantile Origins of Sexual Identity*, op. cit., p. 211. Freud, *Contributions to a Discussion on Masturbation* (1912), *Standard Edition*, vol. 12, pp. 243–54, and his many comments in *Minutes of the Vienna Psychoanalytic Society*, 4 vols, ed. Herman Nunberg and Ernst Federn (New York: International Universities Press, 1964–75). Michael Balint, 'The Adolescent's Fight against Masturbation' (1934), in *Problems of Human Pleasure and Behaviour* (London: Karnac, 1957), pp. 49–68.

p. 74 Effacement, see Louis Linn, 'Some developmental aspects of the body image', *International Journal of Psychoanalysis*, 36 (1955), pp. 36–41. Morris Bender, *Disorders in Perception* (Springfield: Charles Thomas, 1952). On earlier studies of multiple stimulation, see MacDonald Critchley, 'The phenomenon of tactile inattention with special reference to parietal lesions', *Brain*, 72 (1949), pp. 538–61.

p. 75 Scratch, see Amar Dhand and Michael J. Aminoff, 'The neurology of itch', *Brain*, 137 (2014), pp. 313–22, and the debate around STT neurons.

pp. 77–8 Stephen Jay Gould, *The Panda's Thumb* (New York: Norton, 1980). On tools, see Frank R. Wilson, *The Hand*, op. cit.; Raymond Tallis, *The Hand: A Philosophical Inquiry into Human Being* (Edinburgh: Edinburgh University Press, 2003); Colin McGinn, *Prehension: The Hand and the Emergence of Humanity* (Cambridge, Mass.: MIT Press, 2015); and Elaine Scarry, *The Body in Pain*, op. cit., pp. 172–6.

p. 85 On pseudo-activism, see Slavoj Žižek, *Trouble in Paradise* (London: Allen Lane, 2014).

p. 86 Catastrophes, see Richard Kaplan, 'Press, paper and the public sphere', *Media History*, 21 (2015), pp. 42–54.

p. 87 Gift economy, see Walter Burkert, *Creation of the Sacred* (Cambridge, Mass.: Harvard University Press, 1998).

p. 88 Jean Allouch, *Erotique du deuil au temps de la mort sèche* (Paris: EPEL, 1995).

p. 92 King James I, *A Counterblaste to Tobacco* (London, 1604).

p. 94 Beads, see Eithne Wilkins, *The Rose-Garden Game* (London: Gollancz, 1969); Lisa MacKinney, 'Rosaries, Paternosters and devotion to the Virgin in the households of John Baret of Bury St Edmunds', *Parergon*, 24 (2007), pp. 93–114; and Lois Sherr Dubin, *The History of Beads: From 30,000 BC to the Present* (London: Thames and Hudson, 1987).

pp. 96–8 No culture, P. Feyereisen and J.-D. de Lannoy, *Gestures and Speech: Psychological Investigations* (New York: Cambridge University Press, 1991). Ray Birdwhistell, *Kinesics and Context* (Philadelphia: University of Pennsylvania Press, 1970). O. Michael Watson, *Proxemic Behaviour: A Cross-Cultural Study* (The Hague: Mouton, 1970). Paul Ekman and Wallace V. Friesen, 'The repertoire of nonverbal behavior: Categories, origins, usage, and coding', *Semiotica*, 1 (1969), pp. 48–98. Paul Ekman and Wallace V. Friesen, 'Hand movements', *Journal of Communication*, 22 (1972), pp. 353–74. Norbert Freedman and Irving Steingart, 'Kinesic internalization and language construction', *Psychoanalysis and Contemporary Science*, 4 (1975), pp. 355–403. After loss, see George F. Mahl, 'Gestures and Body Movements in Interviews', in J. M. Schlien (ed.), *Research in Psychotherapy*, vol. 3 (Washington: American Psychological Association, 1968), pp. 295–346, and his *Explorations in*

Nonverbal and Vocal Behavior (New Jersey: Lawrence Erlbaum, 1987). Edmund Bergler, *The Writer and Psychoanalysis* (New York: Robert Brunner, 1954). On analytic listening, see Norbert Freedman, 'On psychoanalytic listening: The construction, paralysis and reconstruction of meaning', *Psychoanalysis and Contemporary Thought*, 6 (1983), pp. 405–34. Anna, see 'Memories', *Journal of Child Psychotherapy*, 21 (1995), pp. 375–401; and Elizabeth Young-Bruehl, *Anna Freud* (New York: Summit, 1988), p. 450. Doodling, W. S. Marclay et al, 'Spontaneous drawings as an approach to some problems of psychopathology', *Proceedings of the Royal Society of Medicine*, 31 (1938), pp. 1337–50.

pp. 100–101 Susan Goldin-Meadow, *Hearing Gesture: How Our Hands Help Us Think* (Cambridge, Mass.: Harvard University Press, 2003), p. 143. Writing as embodiment, see Laura Kendrick, *Animating the Letter: The Figurative Embodiment of Writing from Late Antiquity to the Renaissance* (Ohio: Ohio State University Press, 1999). David Efron, *Gesture and Environment* (New York: King's Crown Press, 1941). Weaving, see R. B. Onions, *The Origins of European Thought* (Cambridge: Cambridge University Press, 1951); and Frances Biscoglio, ' "Unspun" heroes: Iconography of the spinning woman in the Middle Ages', *Journal of Medieval and Renaissance Studies*, 25 (1995), pp. 163–76. Jacques Lacan, *Le Sinthome* (1975–6), ed. Jacques-Alain Miller (Paris: Seuil, 2005).

p. 103 Erasmus, *De civilitate*, in J. K. Sowards (ed.), *The Collected Works of Erasmus*', vol. 25 (Toronto: Toronto University Press, 1985), p. 284. Erving Goffman, *Behavior in Public Places*, op. cit.

p. 106 Erving Goffman, *The Presentation of Self in Everyday Life* (New York: Random House, 1959), p. 218.

Acknowledgements

Special thanks to my editor, Simon Prosser, not only for his work on the text but for encouraging me to turn my thoughts into a book. Pat Blackett, Sophie Pathan and Mike Witcombe gave me invaluable help with research, and I would like to thank them and everyone else who has contributed with suggestions, clarifications and corrections: Maria Alvarez, Josh Appignanesi, Chloe Aridjis, Devorah Baum, Susie Boyt, Louise Clarke, Sarah Coward, Elanor Dymott, Astrid Gessert, Antony Gormley, Anouchka Grose, Susan Hiller, Rachel Kneebone, Jeff McMillan, Michael Molnar, Rivane Neuenschwander, Susie Orbach, Cornelia Parker, Vicken Parsons, Ali Smith, Boika Sokolova, Hermione Thompson, Sylvia Whitman and Toby Ziegler. Thanks to Tracy Bohan at Wylie for being a fabulous agent. Mary Horlock's comments were as insightful as ever, though I won't forgive her for making me cut my favourite paragraph about Kentucky Fried Chicken.